DESIGNING LOGIC SYSTEMS USING STATE MACHINES

Camera copy for this book was supplied by the author.
The sponsoring editor was Charles R. Wade, and
editing coordination was managed by Ronald Q. Lewton.
The cover was designed by Judith L. McCarty,
based on materials supplied by the author.
Michael A. Ungersma supervised production.

The book was printed and bound by Shepard's Citations.

**DESIGNING LOGIC SYSTEMS
USING STATE MACHINES**

Printed in the United States of America.

6 7 8 9 0 EBEB 9 8 7 6

07-011120-0

DESIGNING LOGIC SYSTEMS USING STATE MACHINES

Christopher R. Clare
Laboratory Project Manager
Electronics Research Laboratory
Hewlett-Packard Laboratories

McGraw-Hill Book Company
New York San Francisco St. Louis Düsseldorf Johannesburg Kuala Lumpur London
Mexico Montreal New Delhi Panama Rio de Janeiro Singapore Sydney Toronto

FOREWORD

You will find that the material in this book is considerably different from material in previous books discussing logical design. However, it has been around long enough to have been well tested. A dozen years have passed since I discovered the basic design process. Five years ago Christopher Clare began to formalize and expand upon the initial concepts by writing this book. For the past two years a draft of his book has been available within Hewlett-Packard Company and has been used extensively by more than a hundred practicing engineers. The material Chris presents has been used in the design of many successful products including the pocket calculator shown on the cover of this book.

This book introduces an extra step into the logic design process. It is an intermediate notation called an ASM (Algorithms State Machine) chart. The ASM chart separates the conceptual phase of a design from the actual circuit implementation. It gives the designer the freedom of expressing very complex algorithms without worrying about the circuit implementation. Having once created a satisfactory ASM chart for his algorithm, he has the choice of reducing the chart directly into any logic family or implementing it in a read-only-memory.

Our experience shows that some parts of this book will be challenged. Initially you may be bewildered when you see that an ASM looks like a conventional flow chart but is used differently. In an ASM chart the operations described within a box and all branching associated with the box are executed simultaneously, not sequentially as in a flow chart. The reasons for this convention will become evident as you read the book.

I have been told that the entire concept wouldn't work (don't worry, it does) and that it is almost a Mealy representation and that it is almost like a Moore representation and that it is almost like a flow chart. I agree, it is almost many things--but it really isn't any of them. What it is is what this book is about.

Thomas E. Osborne

November 1972

ACKNOWLEDGMENTS

I would like to acknowledge the people who contributed to the development of this book. I would like to thank Kitty Peters for preparing the final copy, Betty Downs and Whitey Pollock for preparing the hundreds of illustrations, and Helen Azadkhanian and Mary Jane Robbins for typing the rough draft.

My special thanks go to Dr. James Duley of HPL/ERL for his careful reading and thoughtful criticisms of all parts of the book. Many sections went through numerous versions to pass his critical eye bent on finding ambiguities.

My thanks to Dr. Paul Stoft, Director of HPL/ERL, for having the confidence in me to provide the time and the atmosphere in which this book was created.

Finally, I would like to acknowledge Tom Osborne for his initial contribution and continued support of the key ideas behind the design process described in this book. The ideas have been employed in HPL/ERL for the past four years with remarkable success. I have only collected these ideas, organized them and expanded them into a general design approach.

Chris Clare

ACKNOWLEDGMENTS

TABLE OF CONTENTS

CHAPTER I
INTRODUCTORY CONCEPTS

1.0 WHERE TO BEGIN

Deciding where to begin must plague every person who decides to describe anything, and this author for one is no exception. In this chapter, the starting point is the BLACK BOX of a logic system. A black box is found by the FUNCTIONAL PARTITIONING of a task which is most suitably described by an ALGORITHM.

1.1 BASIC DESIGN CONCEPTS IN LOGIC

This section introduces the ideas of a logic BLOCK DIAGRAM, of SYSTEM DESIGN and of LOGIC DESIGN. It also introduces three phases in design which are DEFINITION, DESCRIPTION and SYNTHESIS.

An essential concept in logic systems is that the function of a logic machine can be described independently from the details of that machine. This concept leads to the representation of the machine as a black box which behaves in a predictable manner and is represented by a drawing called a BLOCK DIAGRAM. The "behavior" of a machine refers to the relationship between INPUTS and OUTPUTS—shown on the block diagram as TERMINALS—and TIME, which is only implied. Such a basic black box is shown in Figure 1.1.

Figure 1.1 A Basic Black Box of a Block Diagram.

The black boxes that are used in logic systems refer to tasks that can be described by Boolean Algebra. The inputs and the outputs are two-valued; that is, any given input or output can be either TRUE or FALSE. This fact makes possible an exact specification of the black box behavior because there can be no ambiguous inputs or outputs. A logic system is of interest in electronics because the advantages of exact specifications are achieved with simple, reliable circuits.

Two kinds of design, called SYSTEM DESIGN and LOGIC DESIGN, are important in logic systems. The system design defines the black boxes to be used in accomplishing a desired task by specifying the behavior and terminals for each black box which are then called FUNCTIONAL MODULES. Creating a system design in turn depends upon an understanding of logic design, which is the formation of the details in the functional modules.

There are three design areas involved in the process of synthesizing (building) a module, and they are called DEFINITION, DESCRIPTION AND SYNTHESIS. Definition is concerned with the system design and the terminals, description with the details of the logical operations in the modules, and synthesis with the design of actual hardware (circuits) which will execute the description. The discussion of these three design areas constitutes most of this book.

1.2 FUNCTIONAL PARTITIONING

Some system design basics are discussed in this section covering the partitioning of systems into modules and the interconnection of modules with BUSES.

The partitioning of a logic system involves the identification of the basic operational requirements of the desired task and the representation of each by a functional module. In such a way a complex task is divided into smaller constituent tasks that are easy to define. These well-defined tasks become modules which fall into one of three categories, either MEMORY, TRANSFORM or CONTROL. An understanding of these three areas is the first step in partitioning a system into modules.

Module Types

MEMORY modules carry information from one period of time to another. The information STORED in the memory consists of logic values (true or false, 1 or 0) which may be arranged singly, called BITS, or in groups, called BYTES or WORDS. A single memory may hold many bits or words, but for any module there is a limit to the memory, called the STORAGE CAPACITY or more often just capacity for short. Each word of information in a memory is located in a position described by an ADDRESS so that the same information can be found repeatedly.

Memory modules usually handle one address at a time. The address is the input to the memory module, and the corresponding word is the output. A word is said to be READ from memory by such a process. Reading several words takes separate read cycles.

Memory modules may contain fixed or changeable information. Fixed memories store information which may describe the operation of a machine or may represent number tables. These memories are called READ ONLY MEMORIES (ROM) or READ ONLY STORAGE (ROS). Other memories are changeable in that a word found in a particular address can be changed by the inputs and replaced by a new word. The new word is WRITTEN or STORED into the address location. There are two types of changeable memories: REGISTERS and READ-WRITE MEMORIES. Registers can be read and written at the same time while Read-Write memories must be read and written at separate times. Generally, the information read from a memory should be the same as the information written.

TRANSFORM modules produce one set of outputs for each set of inputs, where the important consideration is the logical relationship between the inputs and outputs rather than the time involved in the translation. Some common transform tasks are given special module names. For example, an ENCODER or MULTIPLEXER takes a large number of inputs and converts them to few outputs. A DECODER or DEMULTIPLEXER takes a few inputs and converts them to many outputs. An ADDER takes two coded numbers and produces a sum. A COMPLEMENTER changes all input 1's to 0's and all 0's to 1's. Transform modules can also change the definition of logic levels. A very simple example is a light bulb which changes an electric signal to a visual signal. Such a module (light bulb) is called an INTERFACE from the logic to the operator. It should be clear that the transform modules do most of the work in a logic machine.

CONTROL modules produce outputs which direct the operations of both memory and transform modules. Control outputs depend on the present and the past inputs. Later in this chapter both transform and memory modules are shown to be submachines of a control module.

Functional Division

Modules may be grouped together to simplify the explanation or description of a logic system. For example, it may be reasonable to designate a group of modules collectively as a memory module, even though control and transform modules may be included, if the primary purpose of these modules is to make the memory function in the system. Figure 1.2 shows a group of modules, collectively called a memory but composed of other more basic modules as indicated by the enclosed boxes.

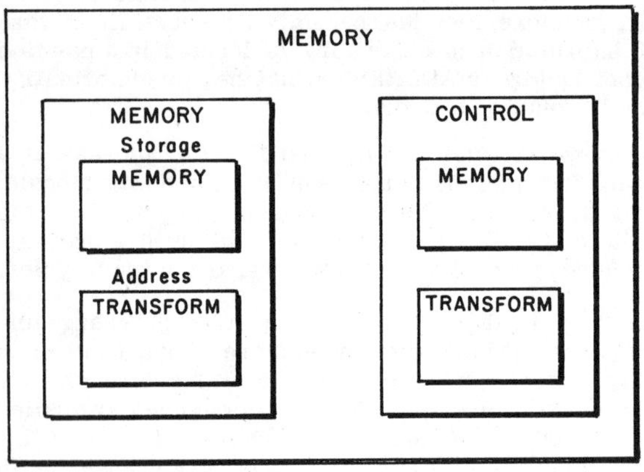

Figure 1.2 A Collective Grouping of Modules Demonstrating Functional Division.

Module Interconnections

A partioned group of modules is connected together by logic-carrying wires called LOGIC LINES. A group of logic lines used in parallel to carry a word of information is called a BUS. The use of interconnection busses is important to the operation of the system.

The BUS STRUCTURE describes the interconnecting logic lines between a group of modules. Figure 1.3 illustrates a basic structure based on a single data bus and one control module. In this structure the communication of data between modules is limited to one word at a time. Additional buses add alternative

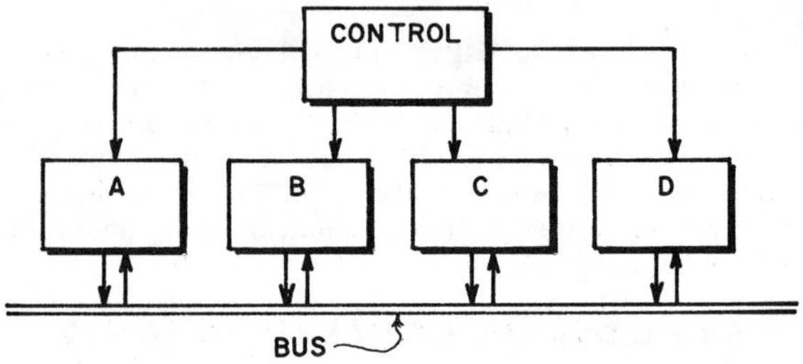

Figure 1.3 A Single Bus Structure.

data paths. In Figure 1.4 all possible bus connections are made. Although this structure is flexible, it is also complicated and expensive. Usually a bus structure is made with as few buses as possible while still providing sufficient speed and flexibility.

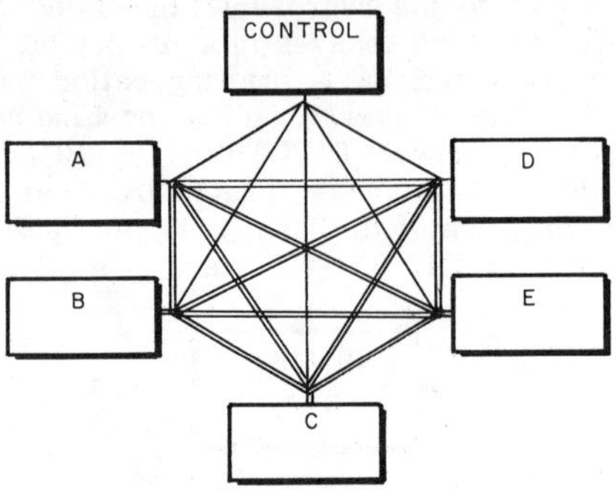

Figure 1.4 A Maximal Interconnect Structure.

Thus functional partioning provides a division of tasks into interconnected modules. What is needed is a concise way of specifying the task and identifying the terminals for each module or the whole system.

1.3 DESCRIBING A TASK WITH AN ALGORITHM

This section describes the algorithm as a process for defining a task. The concepts introduced are the VERBAL ALGORITHM, the OPERATIONAL FLOW CHART and the DESIGN PROCESS.

The fundamental description of the operation of any logic system is called the ALGORITHM. The algorithm is much like a cookbook recipe for producing a result, but is more carefully defined. The following algorithm was given by Euclid in his book, The Elements, and is a means for finding the greatest common divisor of two numbers. It is one of the first historical math algorithms. The algorithm is written in a free form, called a VERBAL ALGORITHM, which consists of the algorithm name, a brief description of its purpose and a listing of its numbered steps. The execution of the steps as described performs the desired task.

ALGORITHM E (Euclid's Algorithm).[1] Given two positive integers m and n, find their greatest common divisor.

E1. (FIND THE REMAINDER) Divide m by n and let r be the remainder ($0 \le r < n$).

E2. (TEST FOR ZERO) Terminate the algorithm if $r=0$ and let n be the answer.

E3. (INTERCHANGE) Replace m with n, and n with r and go to step E1.

Algorithm E could also be represented by an OPERATIONAL FLOW CHART which represents the flow of thought through the algorithm steps as shown in Figure 1.5.

The operational flow chart and the verbal algorithm contain equivalent information. The essential difference in the operational flow chart is that the information describing the operations is separated from the information describing the sequence. Both forms, however, clearly show that the first step is to divide m by n and end the algorithm if the remainder is equal to zero. For a nonzero remainder, m is replaced by n and n is replaced by r. Note that the relative value of m and n is unimportant. If $n>m$, then the first division will have m as the remainder. The next step replaces n with m and r with n, which reverses m and n so that in the next division $m>n$.

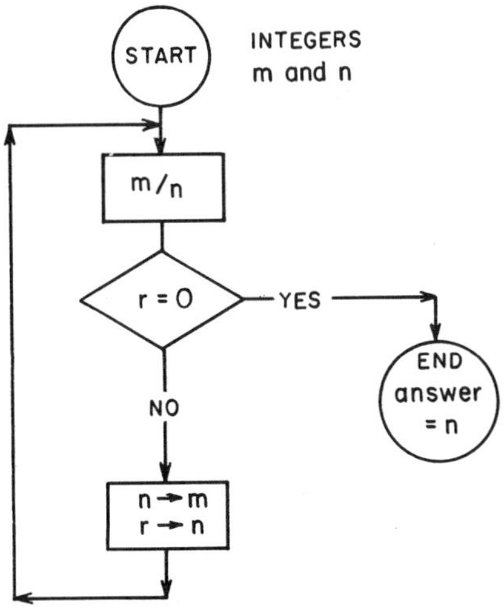

Figure 1.5 Operational Flow Chart for Euclid's Algorithm.

[1] Taken from Don Knuth, "The Art of Computer Programming." See references at end of chapter.

Thus the Algorithm E is a little recipe for finding the answer. Although it could be called a process, a method, a technique or a routine, the word "algorithm" denotes more than just a set of rules which gives a sequence of operations for solving a specific type of problem. An algorithm has five additional important characteristic features:

1. Finiteness. An Algorithm must always terminate after a finite number of steps; that is, the number of steps is countable. In Algorithm E, m is reduced for each loop, so E must eventually terminate.

2. Definiteness. Each step of algorithm must be precisely defined; the actions to be carried out must be rigorously and unambiguously specified. Algorithm E requires positive integers because step E1, divide m by n, is not defined, for example with $5/\pi$ or $3/0$.

3. Input(s). An Algorithm may or may not require inputs, which are initial quantities given to it before the algorithm begins. In Algorithm E, for example, m and n are inputs.

4. Output(s). An algorithm has one or more outputs specified in the description which are quantities related to the inputs or characteristic to the algorithm. Algorithm E has one output in step E2, which is the greatest common divisor or the two inputs.

5. Effectiveness. An algorithm is expected to produce a useful result. That is, the outcome of the algorithm yields a predictable result which has an application. Algorithm E satisfies this requirement.

The process of designing logic may be described by the flow chart in Figure 1.6 containing the following five phases of design.

1. Definition. Select a set of well defined hardware functions which should do the job.

2. Description. Describe an algorithm for using the hardware defined in phase one to perform the desired task.

3. Evaluation. Evaluate the operation of the definition and description. If the performance is unsatisfactory, change the definition of description until satisfactory performance is achieved.

4. Synthesis. Translate the algorithm description into hardware.

5. Test. Perform checks on the hardware to verify operation as described in the algorithm.

The first loop in the flow chart represents the area of greatest probable activity. Once the evaluation phase has been passed, the design is well on its way to completion. The description phase produces the details of the machine algorithm and depends mostly upon the spark of invention. The quality of a logical design relies on the cleverness of the algorithm used to describe the task.

In later chapters, each of the five phases of the logical design process is developed in terms of a set of practical procedures based on a firm algorithmic machine structure. In the next chapter the details of defination are introduced. The description languages are covered in Chapter 3, synthesis in Chapters 4 through 6, and evaluation and test in Chapter 7.

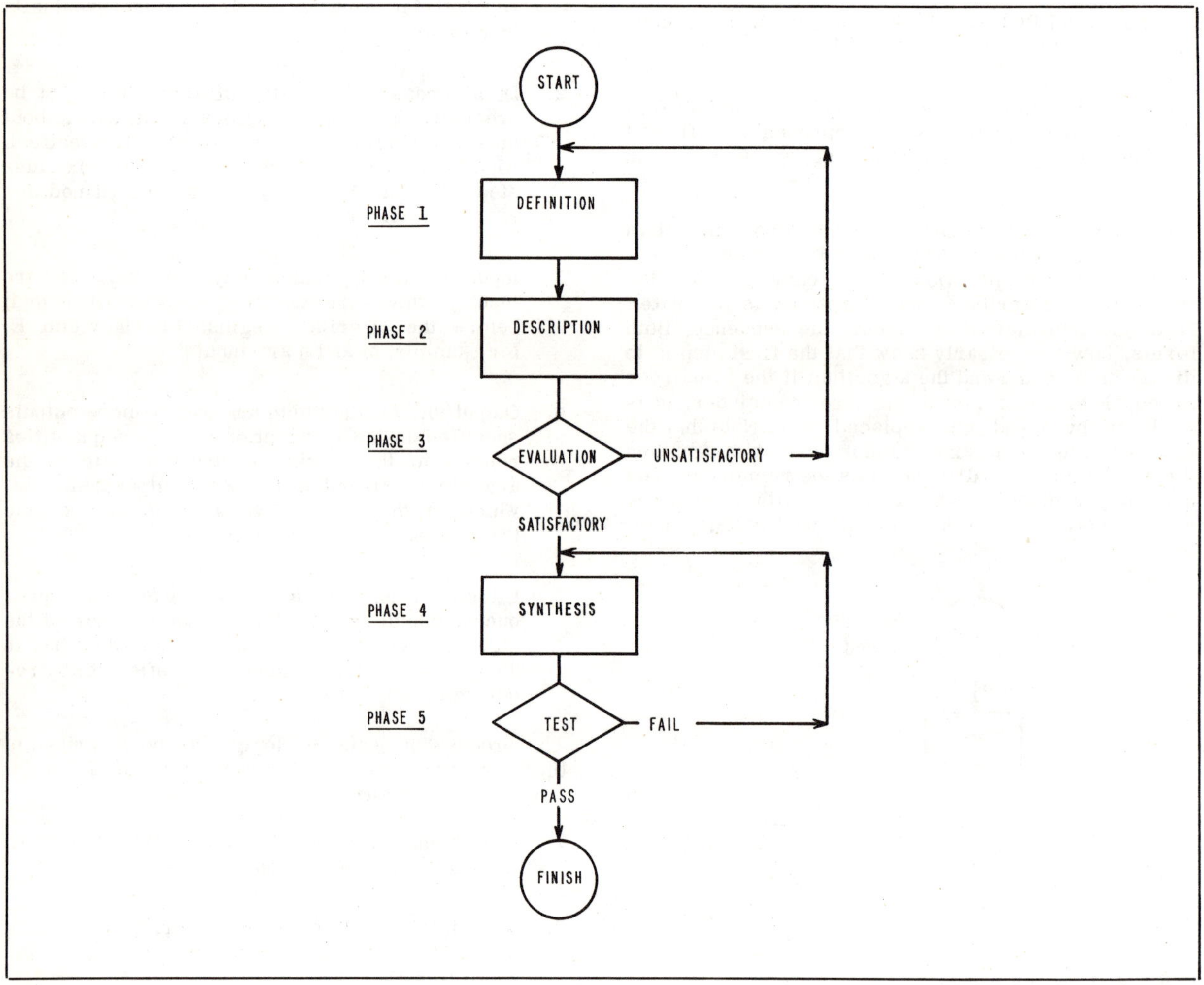

Figure 1.6 Flow Chart of The Five Phases of the Design Process.

REFERENCES
CHAPTER I

MACHINE ORGANIZATION

Flores, Ivan: Computer Organization (Englewood Cliffs, New Jersey: Prentice-Hall, Inc., 1969).

Hellerman, Herbert: Digital Computer System Principles (New York: McGraw-Hill Book Company, 1967), Chapter 5.

Wegner, Peter: Programing Languages, Information Structures and Machine Organization (New York: McGraw-Hill Book Company, 1968). Chapter 1.

ALGORITHM

Knuth, Donald E.: The Art of Computer Programming, Vol. 1: Fundamental Algorithms (Menlo Park, California: Addison-Wesley Publishing Company, Inc., 1968), Chapter 1.

CHAPTER II

THE MODULE DEFINITION

2.0 THE DEFINITION PHASE

Modules can be defined in terms of the terminals used as inputs and outputs. This chapter clarifies the definition of these terminals by proposing a general STATE MACHINE model for all the modules and a system of MNEMONICS to use on the terminals. A number of useful concepts are brought together in a sample definition developed in the last section.

2.1 THE STATE MACHINE

Every module of a logic system can be represented by a general model called the STATE MACHINE. This model contains the elements required to describe the module behavior in terms of its inputs, outputs and time. Figure 2.1 shows three elements in the general model: the NEXT-STATE FUNCTION, the STATE and the OUTPUT FUNCTION. The inputs and the outputs, which pass through the module to the outside world and to other modules, are also called QUALIFIERS and INSTRUCTIONS as indicated. These elements are explained further.

The STATE of a machine is the memory of sufficient past history to determine future behavior. In terms of the state machine, this means sufficient information to determine both an output and a next state if the present inputs are known. In the machine, such a memory is usually made from bistable circuits called FLIP-FLOPS. A group of flip-flops forming the state is called the STATE REGISTER. A different state is defined for each combination of stored bits, which means that there are 2^n possible states for n state register flip-flops.

[1]The words "next-state" form a compound adjective, as opposed to "next state", which is an adjective modifying a noun.

The flip-flops of the state are called STATE VARIABLES and are defined by a STATE DECLARATION. Each variable is given a name such as A, B or FF6. The group of variables making up the state is then tied together by a CONCATENATION OPERATION, $\frown$, between each variable of the state. For example, if A, B, C, and D are the variable names, then STATE = $D\frown C\frown B\frown A$ would be a form of the state declaration. If A = 1, B = 0, C = 1 and D = 1 the particular state can be represented by the code 1101, which places the logic value for each variable in the position corresponding to the state declaration. Other registers can be defined in this same way.

Each state of a machine has a next state determined by the NEXT-STATE FUNCTION. The STATE TIME is normally determined by a periodic input to the state register. At the end of each state time the next state becomes the present state. The next-state function, g, depends upon the present state, X, and the inputs or qualifiers, Q. If the basic state time is represented by T, and k is a counting integer, then X(kT) represents the state at the discrete time kT. Using this terminology, the next-state function, g, can be defined as follows:

$$X\big((k+1)\,T\big) = g[X(kT),\ Q(kT)]$$

The notation for the next-state operation is simplified by using the DELAY OPERATOR which is an arrow $\rightarrow$ or $\leftarrow$, pointing in the direction of the next-state replacement. The next-state function is now written $X \leftarrow g[X, Q]$ which means that the value of X is changed at the end of a state time and the new value is given by g [X, Q] where X and Q are values during the present state time. Since the changing of the state is delayed from the determination of the next state, the operator is called the delay operator.

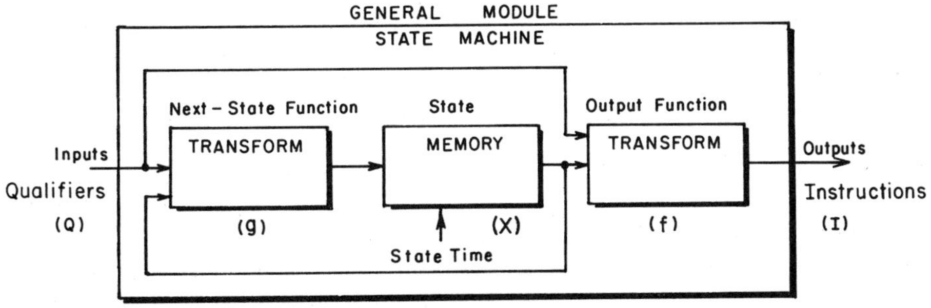

Figure 2.1 The General State Machine Module.

The OUTPUT FUNCTION generates a set of outputs or INSTRUCTIONS, I, from the state and input information for each state. Like the next-state function, it consists of a transform operation, called f, which has the following expression:

$$I(kT) = f[X(kT),\ Q(kT)]$$

The k and T are the counting integer and basic time, respectively, as in the next-state function. The notation for the output function may also be simplified by defining the equal sign, =, as an IMMEDIATE OPERATOR which represents an operation in the present state time. Accordingly, the terminology for the output function is simplified to

$$I = f[X, Q]$$

The operation of a state machine cycles in an orderly manner, reaching a stable condition during every state time, kT. The state and, consequently, the next state and the outputs are defined only during the stable period of the state time. Figure 2.2 illustrates the division of the state time into a transition period followed by a stable period. The transition period is determined by circuit delays.[1] The length of the stable period is the difference between the state time and the transition time, and this length of time must be greater than zero for the state machine to be defined. Therefore, the state time must be greater than the transition time. The operation of the state machine can be visualized as a series of steps consisting of the outputs in each stable time. For example, in Figure 2.3 the kT time notation is used to represent the changes and relationships in the operations during the stable period for three successive state times.

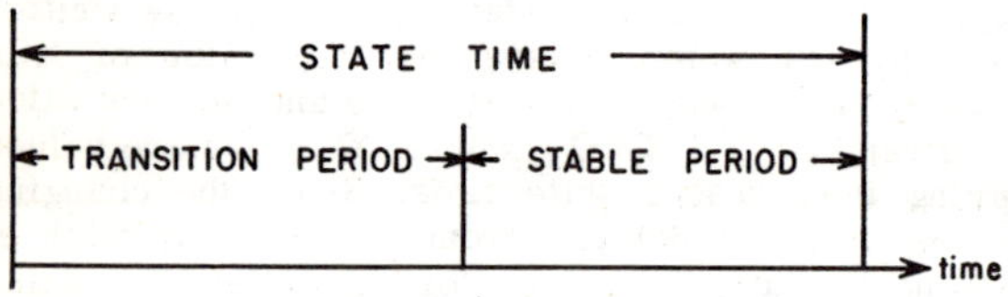

Figure 2.2 The Two Periods in a State Time.

STATE	X(0)	X(T)	X(2T)
INPUTS	Q(0)	Q(T)	Q(2T)
NEXT STATE	g[X(0), Q(0)]	g[X(T), Q(T)]	g[X(2T), Q(2T)]
OUTPUT	f[X(0), Q(0)]	f[X(T), Q(T)]	f[X(2T), Q(2T)]
	0	1T	2T — 3T time

Figure 2.3 Time-Line Representation of the State Machine Functions.

[1] Circuit delays correspond to the "settling time" of gates. See Section 3.5 for further discussion of these times.

In the next chapter, specific methods for describing the next-state function and the output function are covered. This chapter continues with the definition of terminals.

2.2 TERMINAL DEFINITION

This section describes a system of MNEMONICS for defining input and output terminals and some semantics for defining the meaning associated with each terminal.

A System of Mnemonics

The terminals forming the inputs and outputs of the state machine are given names, called MNEMONICS, consisting of letters and numbers, which are used as memory aids in recalling the functions of operations associated with the logic levels on the terminals. These names are short, making them easy to manipulate in the design process. Although the choice of a naming system is personal, the system chosen for this book is recommended. This system fulfills four objectives, which are to provide (1) a single name for each common logic line, (2) a consistent logic level throughout the machine, (3) an indication of the terminal type (input, output), and (4) a means of identifying the logic interpretation.

The basic mnemonic is a group of three or four letters which is usually formed from the first letters of the words describing the operation or function of the related terminal. For example, a terminal used to reset a register might be called RST. The symbols used are chosen from an ordinary typewriter set to ease the job of documentation.

An initial letter is used to signify the logic level and the type (input or output) of terminal. For example, RST is an output which may reset the register when RST is a 1 or when it is a 0. A prefix of L (LRST) signifies that the reset occurs on a 0 (LOW) signal, while a prefix of H (HRST) signifies that the reset occurs on a 1 (HIGH) signal. The logic level of 0 is the less positive or LOW level. The inputs are designated by the prefix Y (YES) if the test is true on logic level 1 or N (NO) if the test is true on the logic level 0. An input terminal testing the count of 10 in a counter, with the basic mnemonic of CET (Count Equal Ten), can be named YCET or NCET, depending on the logic level when the count equals ten.

While H, L, Y and N all involve the logic level of the terminals and are therefore inportant in the synthesis phase of design, another prefix, I, is useful in the interpretation of the output or instruction terminals. I stands for immediate function, which means a task that is completed in the present state time. H or L without a prefix indicates a delay function, which means a task that is completed in the next state time. The use of I, then, is related to the operations in another module which become important to the understanding of the module being designed. Figure 2.4 summarizes the interpretation of the basic prefix letters.

TYPE OF LOGIC TERMINAL	MNEMONIC INITIAL LETTER	MEANING FOR LOGIC TERMINAL EQUAL TO	
		1	0
OUTPUT OR INSTRUCTION	H	operation performed	inactive
	L	inactive	operation performed
INPUTS OR QUALIFIERS	Y	statement is true	statement is false
	N	statement is false	statement is true
IMMEDIATE	I	precedes either H or L to designate an immediate function	

Figure 2.4 Summary of Prefix Letters.

Terminal Interpretation

The terminal definition is completed by specifying the exact meaning of each type of terminal, using some simple semantics. The outputs are always immediate functions as far as the module being designed; however, the output may produce an immediate or a delayed response in another module. The external response is indicated in a statement defining the terminal by enclosing it in square brackets. For example, if IHRST is an immediate reset instruction which sets a register R to 0, then the definition would look like

IHRST = [Reg. R = 0]

If HRST is a delay instruction to do the same function in the next state, its definition would look like

HRST = [0→Reg. R]

where the bracketed operation is a delay function. It may seem that the use of I is redundant but, later on, the mnemonics are used alone and the I is then useful. The inputs are defined by a simple equivalence between the mnemonic and the meaning. For example, a test for overflow of a memory might be

YOFL = memory overflow

2.3 A SAMPLE DEFINITION

This section ties together the ideas of functional partitioning and terminal definition by defining a simple machine to act as the dealer for the Black Jack or "21" card game. The complexity of this machine is suitable for illustrating the definition process while being neither trivial nor overwhelming. In later chapters, the Black Jack machine is again used to demonstrate other design phases.

The Black Jack machine is described by Algorithm B, which summarizes the basic rules involved in the game and describes the process as seen by the game dealer rather than the player.

ALGORITHM B (Black Jack Dealer). Given that cards are valued 2 through 10 and aces are valued 1 or 11, play the "dealer's" logic to produce a "stand" or "broke" game completion.

B1. (START A NEW GAME) Set score to zero and reset the 11-point ace flag.

B2. (MAKE A HIT) Accept a card and add its face value to score.

B3. (CHECK FOR AN ACE) Add 10 more to score if the card entered is an ace and the 11-point ace flag is reset. Set the 11-point ace flag.

B4. (CHECK FOR A HIT) Go to B2 if score is 16 or less.

B5. (CHECK FOR A STAND) Indicate a stand and go to B1 if the score is 21 or less.

B6. (SUBTRACT AN ACE) Subtract 10 from score, reset the 11-point ace flag and go to B4 if the 11-point ace flag is set.

B7. (INDICATE A BROKE) Indicate a broke and go to B1.

The machine might look something like Figure 2.5. The machine has a card reader to read the value of the card. Three lights indicate "stand," "broke" or "hit." A new game is begun with the card following the broke or stand light, which corresponds to returning to step B1 in Algorithm B.

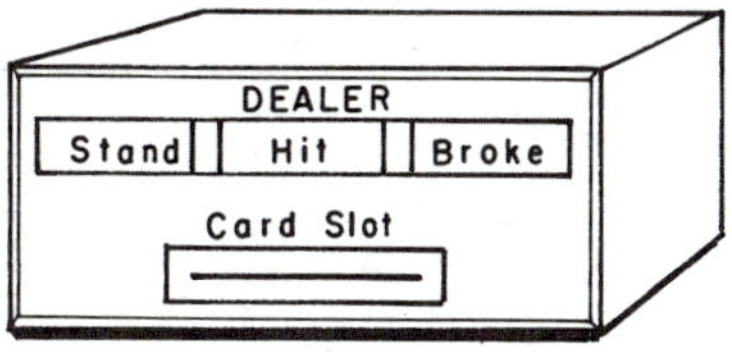

Figure 2.5 Black Jack Dealer

Algorithm B can also be described with the flow chart given in Figure 2.6. At this point in the design, when the algorithm is expressed as a verbal description of operations performed, the flow chart will be called an OPERATIONAL FLOW CHART. Each box will correspond to some operation which may not correspond to any particular hardware.

The operational flow chart or the verbal algorithm points out specific operations which must be accomplished by a logic machine implementing the algorithm. These operations can be divided into several distinguishable modules. There will be memory modules to remember the score and the 11-point ace. There will be several transform modules, one to read the card, one to light the lights, and one to add new cards to the score. The adder module can also subtract the 10 points in step B6 by adding a -10. There must be some means to sequence these operations to perform the desired job, and this function is provided by the control module. Lastly, there must be some bus structure to handle the card value as it goes from the reader to the adder and then to the score register. In this design definition phase, each module is identified and a set of mnemonics is chosen for all the functions and terminals required.

In the following discussion, signals are named for each module of the Black Jack machine. Assume that the card value read by the card reader is transferred to a register called the CARD REGISTER. An instruction to transfer this value will be called:

HTVC = [Card register ← card value]

Assume also that the two values required to handle the additional conditions imposed by an ace are also transferred into this register by two more operations, which will be called:

HT10 = [Card register ← decimal 10]

HT22 = [Card register ← decimal -10]

A qualifier is required to detect an ace in the card register. This qualifier will be called:

YACE = an ace is in the card register

An additional signal will be required to prevent more than one card value from being added after each card entry. This entry process may be described by Algorithm A as follows:

ALGORITHM A (accept a card). Using the card's presence, read a single value from each card.

A1. (WAIT) Wait until the card is present.

A2. (READ) Read the card and put its face value into the card register.

A3. (WAIT) Wait until the card is removed then, go to A1.

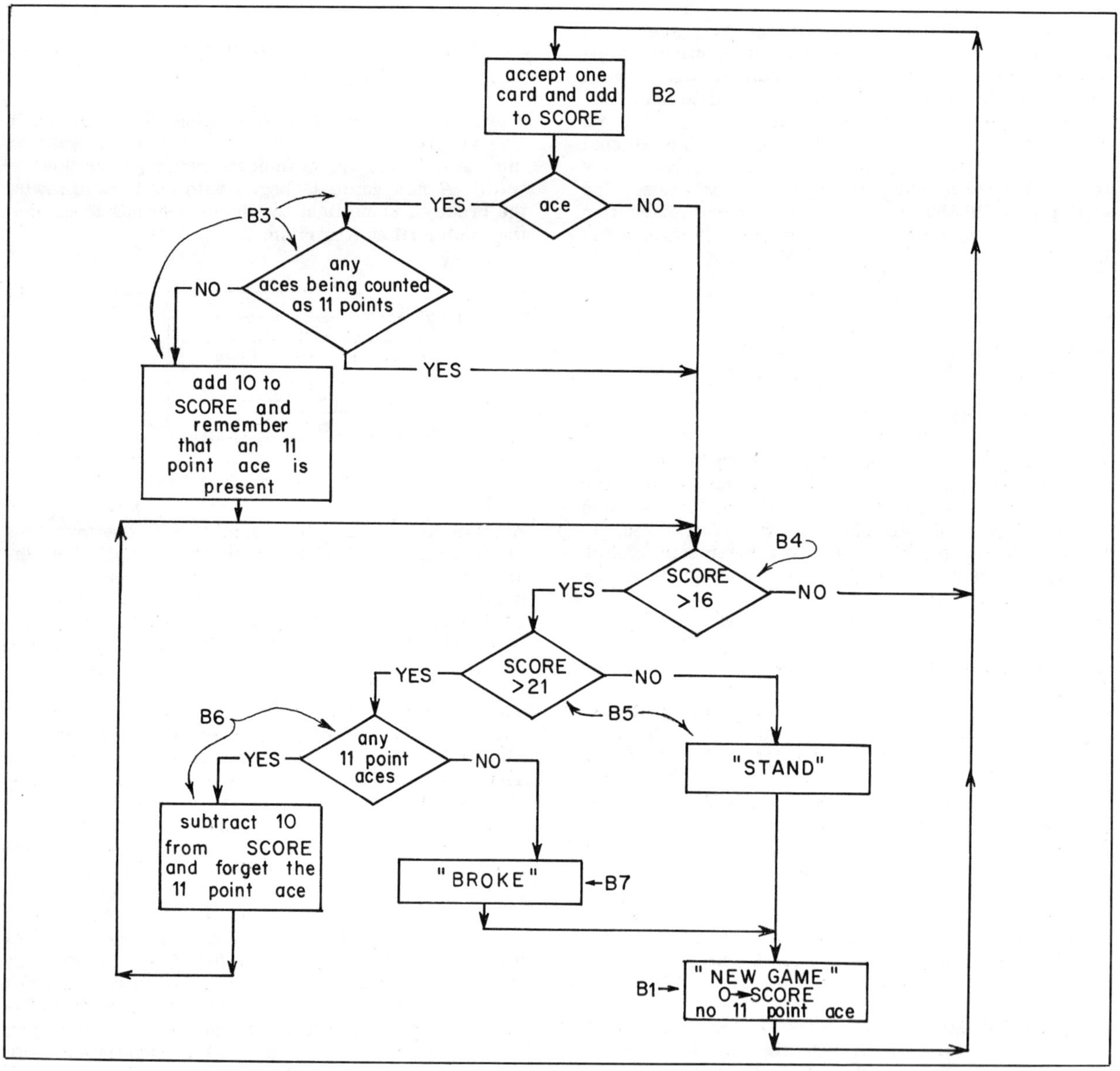

Figure 2.6 Operational Flow Chart for the Black Jack Dealer Machine.

An additional terminal to tell when the card is present will be called:

YCRD = a card is present

The adder has only one control, which will be

HADD = [sum register card register (+) sum register]

(Note: The symbol (+) means arithmetic sum, as opposed to the logic cross, +.)

The sum register has one control to clear the register, that is, initialize it to zero. This terminal will be called:

HCLS = [sum register←0]

Two qualifiers are required to determine the game status. These terminals will be called:

YG21 = the sum is greater than 21

YG16 = the sum is greater than 16

The flag to remember the 11-point ace requires two signals to set and reset it. These terminals will be called:

HJ50 = [11-point flag←1]

HK50 = [11-point flag←0]

YF50 = an 11-point ace has been counted

The lights require one terminal per indication. These terminals will be called:

IHHIT = [HIT light = ON]

IHSTD = [STAND light = ON]

IHBRK = [BROKE light = ON]

All the terminals are drawn with the modules to form a block diagram summary of the terminals and their connections as shown in Figure 2.7.

This diagram, along with the definitions for each mnemonic, serves to demonstrate that a set of operations can be defined to perform a verbal algorithm without the specification of any circuit details. This definition of the basic system forms the first of the five major design phases, which are definition, description, analysis, synthesis and test. As the design progresses, some of the original definitions may be changed or eliminated. However, most of the terminals will appear in the final design. Thus, the definition is an initial foundation for design.

2.4 DEFINITION DOCUMENTATION

The set of terminal definitions forms the first part of a complete set of design documentation which will eventually describe the design philosophy for each logic module. The ultimate goal in a logic design is to produce both a circuit which does the job and a set of documentation which fully describes its logic operation. No logic design is complete without this documentation. Its absence can only mean that the design is incomplete or misunderstood. Logic design is primarily a symbolic manipulation process. It can be done without a systematic approach no easier than two 10-digit numbers can be multiplied in the head. There is entirely too much information to be remembered for an undocumented job to be totally successful.

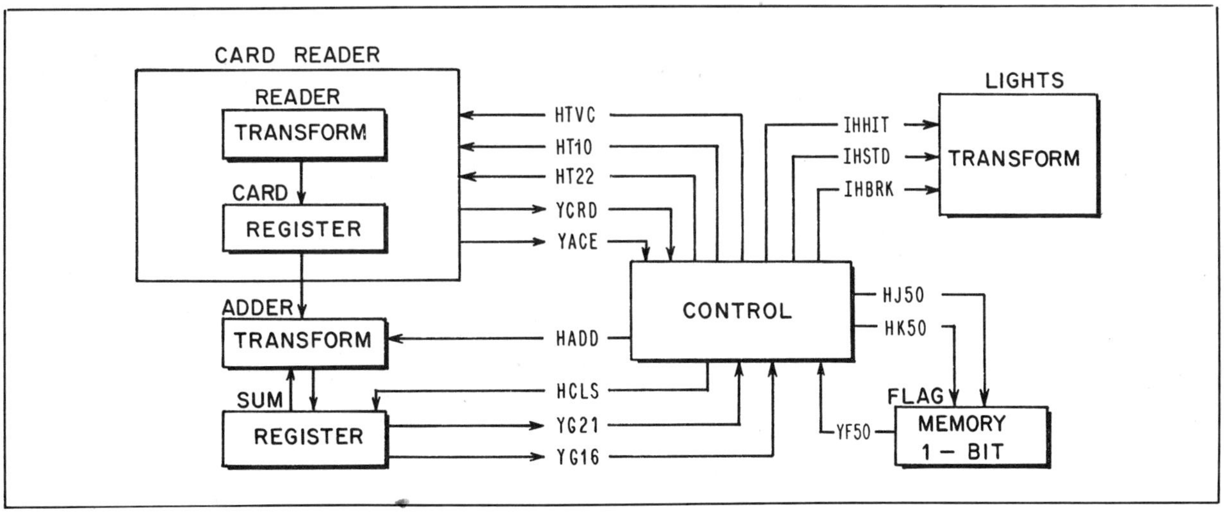

Figure 2.7 A Block Diagram Definition for the Black Jack Machine.

REFERENCES
CHAPTER II

CONCEPT OF STATE

Bertram, J.E.: "The Concept of State in the Analysis of Discrete Time Control Systems," 1962 Joint Autom. Control Conf. New York University (June 27-29), Paper No. 11-1.

Derusso, Roy, and Close: State Variables for Engineers (New York: John Wiley and Sons, Inc., 1967), Chapter 6.

Moore, E.F.: Gedanken-Experiments on Sequential Machines (Princeton, New Jersey: Automata Studies, Princeton University Press, 1956), pp. 129-153.

CHAPTER III

MACHINE CLASS DESCRIPTIONS

3.0 THE DESCRIPTION PHASE

The four basic languages for describing the logical operation of state machine modules are introduced in this chapter. These four languages are the BOOLEAN EXPRESSION, the TRUTH TABLE, the MAP and the ASM CHART. ASM is short for ALGORITHMIC STATE MACHINE. The ASM chart is important because it describes both an algorithm and a state machine simultaneously. These description languages are elaborated upon in the course of describing five classes of state machines, all of which are portions of the state machine model described in Chapter 2.

3.1 THE THREE CLASSIC DESCRIPTION LANGUAGES OF LOGIC

This section introduces BOOLEAN EXPRESSIONS, TRUTH TABLES and MAPS as languages for describing a logic expression. A more complete discussion is given in reference at the end of Chapter 4.

Logic and Boolean Expressions

Logic is an everyday reality in the world around us. The concepts of collective groups denoted by the word "and," or conditional statements beginning with the words "if," "while," "for," and "when," or alternative statements using the words "then," "else," and "or," or negative statements using the word "not" are all examples of words in our everyday language that relate to logic just like this sentence which is itself a list of alternatives separated by the word "or." Logic statements are characterized by a forcefulness and clarity as demonstrated by these well-known phrases: "To be or not to be," "Do or die," "Don't shoot until you see the whites of their eyes," "If I've said it once, I've said it a thousand times."

In a more positive sense, logic can be used as a tool in the study of deductive reasoning, also known as propositional logic. One of the most significant advances in logic was made in 1854 when Boole postulated that symbols could represent the structure of logical thought. His symbolic logic language made possible great advances in the clarity, understanding and documentation of the process of logical thought. The basic symbols are a 1 for TRUE and a 0 for FALSE. There are also many symbols used to represent the relationships between the facts of an expression. The assertions or facts are represented by letters or symbol groups. For example, the statement "When the runners are ready and the gun is fired, the race is started," can be divided into three assertions represented by the following letters:

A = runners are ready

B = gun is fired

C = race is started

Using a raised dot $\cdot$, for the relationship AND, the statement may be represented by the expressions:

A AND B EQUALS C

or $A \cdot B = C$

From this statement it can be seen that C is only true when both A and B are true, which can be said in symbols, $1 \cdot 1 = 1$. Basic relationships, such as this one, are called POSTULATES. The other basic relationships in logic are the OR, represented by a cross, +, and the COMPLEMENT, represented by an overbar, $\overline{A}$, which means NOT A. The set of symbols used in this book is shown in Figure 3.1, and the basic postulates for both AND and OR are shown in Figure 3.2. These relationships form the basis for BOOLEAN ALGEBRA, which is the study of the relationships of two-valued variables such as the assertions discussed above. The use of Boolean algebra is fully described in Chapter 4. Here the primary concern is the representation of a task with symbols.

OPERATION	SYMBOL	EXAMPLE	RESULT
LOGIC OR	+	A + B	1 or 0
LOGIC AND	●	A ● B	1 or 0
EQUAL	=	A = B	Equal
LOGIC COMPLEMENT	$\overline{X}$	$\overline{A}, \overline{B}$	1 or 0
MATH ADD	(+)	A(+)B	Number
MATH MULTIPLY	(●)	A(●)B	Number
DELAY OPERATOR	⇄	A ← B B → A	A takes value of B
MATH SUBTRACT	(−)	A(−)B	Number

Figure 3.1 The Set of Logic Symbols Used.[1]

$X = 0$ if $X \neq 1$	$X = 1$ if $X \neq 0$
$\overline{0} = 1$	$\overline{1} = 0$
$0 \cdot 0 = 0$	$1 + 1 = 1$
$1 \cdot 1 = 1$	$0 + 0 = 0$
$1 \cdot 0 = 0 \cdot 1 = 0$	$0 + 1 = 1 + 0 = 1$

Figure 3.2 Basic Postulates for Boolean Expressions.

[1] A variety of other symbols are also in common use. For instance, the logic OR is often U or V; the logic AND is often ∩, Λ or no symbol; and the logic negation for A is often -A or A'.

Tabular Descriptions

A table is a convenient language for describing all the relationships and outcomes in an expression. Figure 3.3 gives a table to describe the A · B = C expression which was previously used to describe the start of a running race. This table describes the AND function between any two variables, A and B. Figure 3.4 gives tables describing the OR function and the COMPLEMENT function. In each example, C is the result of the operation. Thus, the truth table forms a second language to describe logic relations.

1 = TRUE, 0 = FALSE

LOGIC VALUE OF A	LOGIC VALUE OF B	LOGIC VALUE OF C
0	0	0
0	1	0
1	0	0
1	1	1

Figure 3.3 Truth Table Describing A · B = C.

LOGIC OR

A	B	C
0	0	0
0	1	1
1	0	1
1	1	1

C = A + B

COMPLEMENT

A	C
0	1
1	0

$C = \overline{A}$

Figure 3.4 Truth Tables for the OR and Complement Functions.

The Karnaugh Map Description

The KARNAUGH MAP description is a special form of a table arranged in such a way that the entries representing binary codes (made from two-valued variables) that differ in only one variable are clearly shown by their position. The map is derived from visualizing the code as a vector with n variables representing a point in n-space. The map is a two-dimensional representation of this space. The adjacent codes are called UNIT DISTANCE codes. The Karnaugh map, hereafter called a map, has particular significance in visualizing the simplest expressions, a subject which is covered in Chapter 4. The formation of maps; however, requires further explanation here.

A map can be drawn using an unfolding process which lays out the n-dimensional space into a well-described two-dimensional plot. Each square on the plot will correspond to one code or vector, and information written in that square will correspond to some characteristic value or quality associated with that vector. Figure 3.5 describes the translation of a three-dimensional cube, formed by a three-dimensional vector [C⌢B⌢A], into a map with labeled fields A, B and C. The three-dimensional cube is called a 3-cube for short. In the map, all the squares lying within a field designated by one variable have a code with that variable equal to 1. Outside of the chosen field, the same variable is equal to 0. For example, the three-variable map for the 3-cube has C designating the field of the four lower squares. In each of these squares C = 1. On the other hand, in the upper four squares, C = 0. A designates the center 4 squares. All the squares in this field have A = 1. Because of the way in which the 3-cube is unfolded, the codes for the squares above, below, to the left, or to the right of any square are a unit

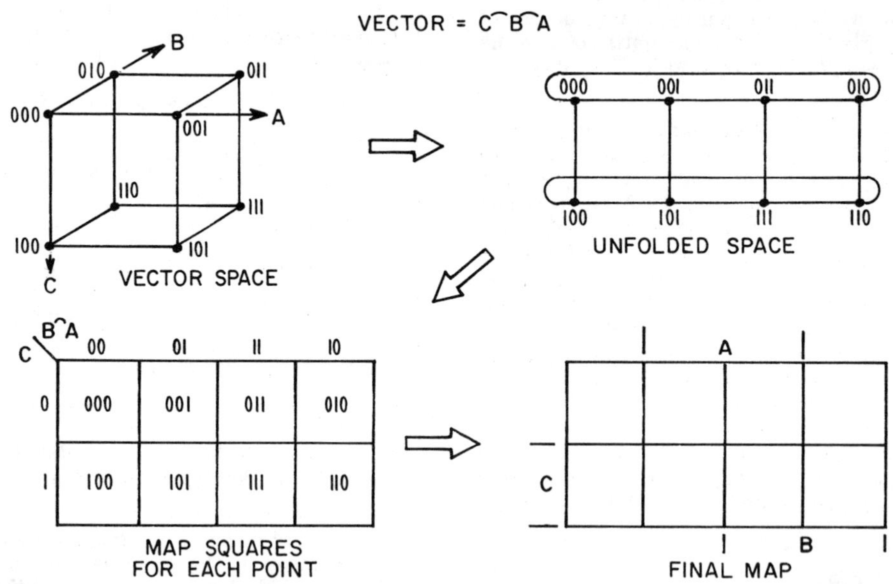

Figure 3.5 The Formation of a Three-Variable Map from a Three-Variable Cube (3-Cube).

distance from the code in that square including all the squares on the ends because the map represents a cube unfolded such that the leftmost column is a unit distance from the rightmost column. Figure 3.6 summarizes the map representations of the remaining n-cubes up to a 6-cube. The 5-cube and the 6-cube are a little different from the smaller maps in that they are really two and four copies, respectively, of a 4-cube map. This choice is primarily for convenience.

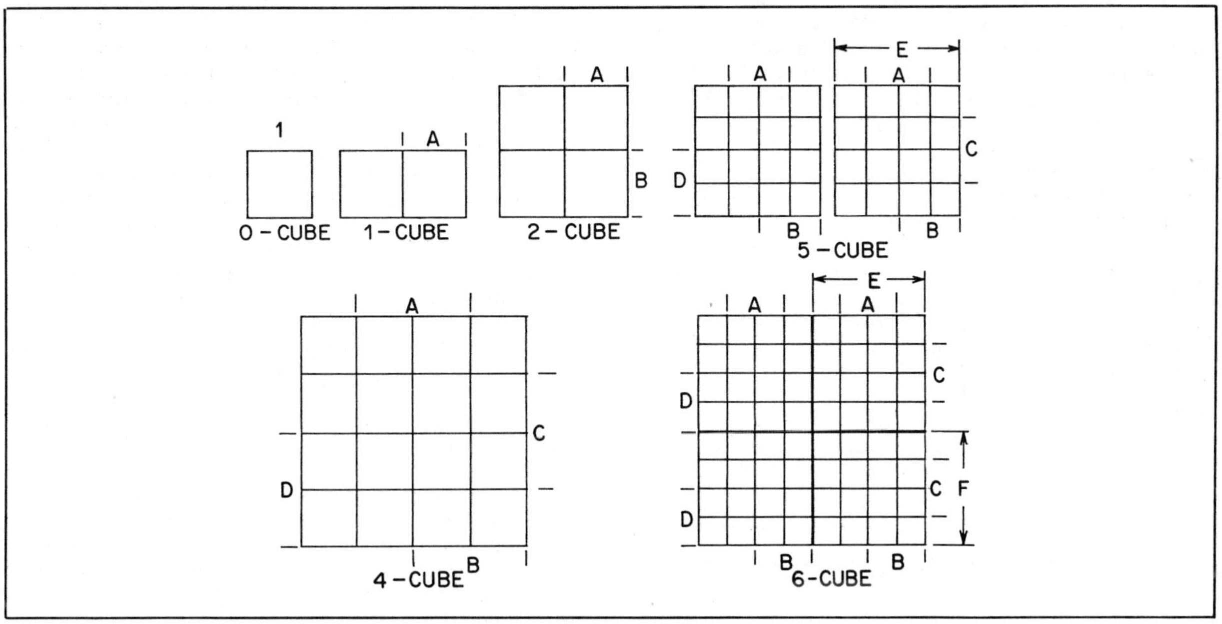

Figure 3.6 Example Map Representations of Various n-Cubes.

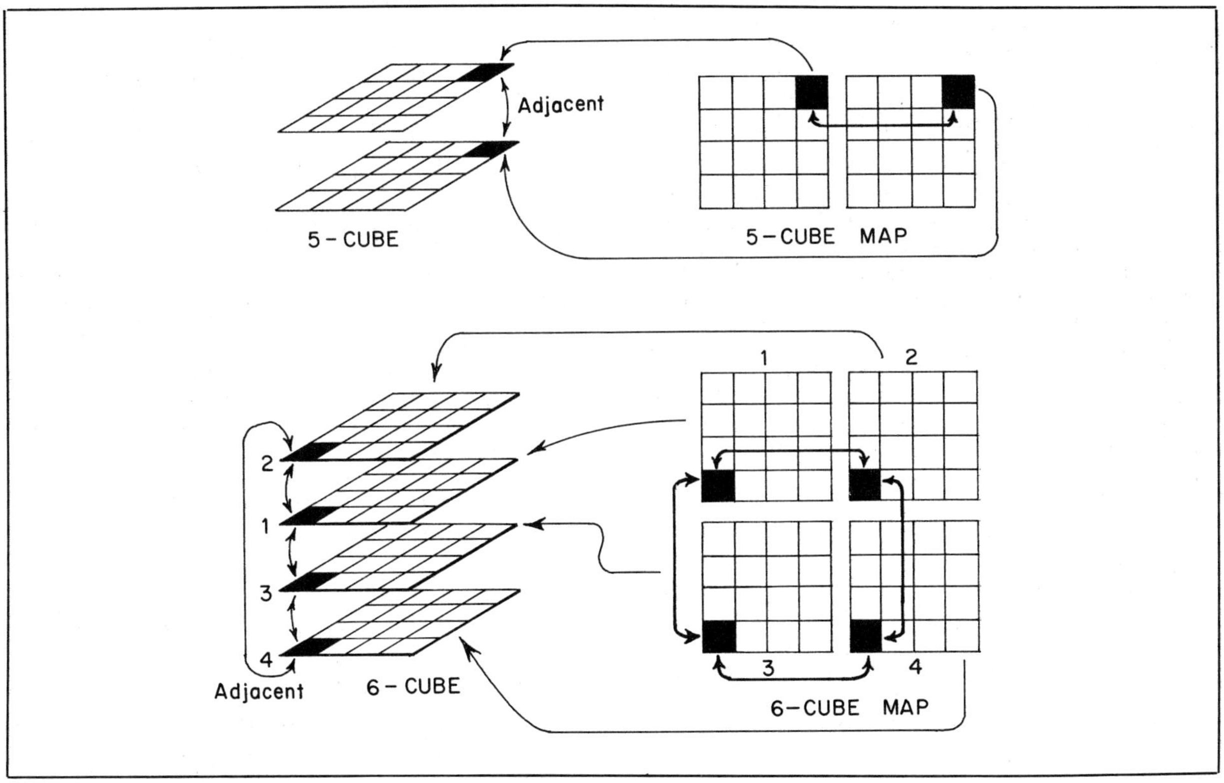

Figure 3.7 The 5-Cube and 6-Cube Map Relationships.

To understand the unit distance or adjacency structure in five and six variable maps, they are viewed as stacked 4-cubes in Figure 3.7. In the 5-cube map, the left and right halves are adjacent. The six-variable half map such that the upper right corners of the left and right halves are adjacent as indicated. In a similar manner all the corresponding squares on the left and right halves are adjacent. The six-variable map adjacency is a simple extension of a five-variable map adjacency as indicated in Figure 3.7 for typical corresponding squares.

Sample map descriptions for the AND, NOT and OR functions are given in Figure 3.8. Thus, maps are yet another way to describe relationships between logic variables.

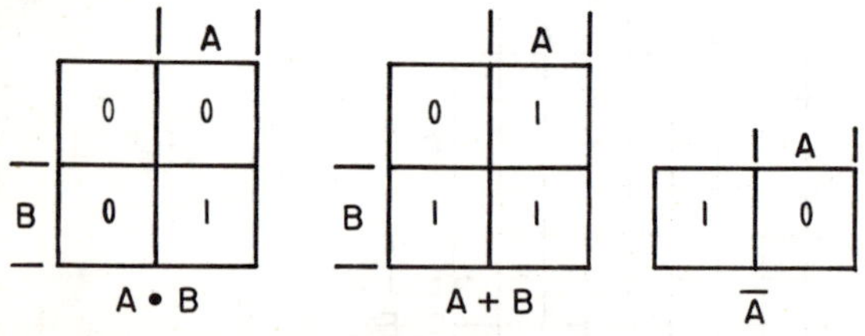

Figure 3.8 Maps for the Three Basic Functions.

3.2 THE ASM CHART DESCRIPTION OF LOGIC

This section describes the basic symbols used in forming an ASM (Algorithmic State Machine) chart. The ASM chart is a diagrammatic description of the output function and the next-state function of a state machine. It is used as an aid in designing a state machine to implement an algorithm and becomes part of the design documentation when completed. The symbols covered are the STATE BOX, the DECISION BOX, the CONDITIONAL OUTPUT BOX and the ASM BLOCK.

The ASM chart has three basic elements: the state, the qualifier and the conditional output. A single state is indicated by a STATE BOX which contains a list of state outputs as shown in Figure 3.9. The state has a name, letter or number, encircled on the left or right of the state box. The state code can be placed along the upper edge of the state box. The state name can be used for referencing the state without knowing the exact state code. The code refers to a particular solution and is probably unknown when first drawing the ASM chart.[2]

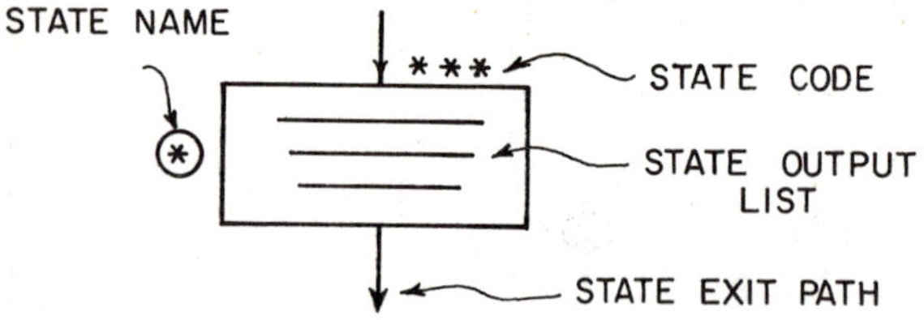

Figure 3.9 The State Box for a Single State.

[2] The state code is given to the state during state assignment, (see Section 4.8).

The OUTPUT LIST consists of mnemonics selected from a defined set of operations. The mnemonics name outputs which are given during the state time. The effects of these outputs are divided into immediate operations, which are defined by an expression using the = symbol, and delay operations, which are defined by expressions using the ← or → symbol. In interpreting an ASM chart, it is important to know which operations are immediate and which are delayed. Usually fewer of the operations are immediate; consequently, to save writing, the immediate instructions are marked with an initial letter I for Immediate, while the delay operations are unmarked as was introduced in Chapter 2. For example, in an output list the immediate instruction, LRST, would be written ILRST while the delay instruction, HJ50, would have no I prefix. The immediate and delay operations are further clarified in Section 3.5. The exit path of the state box leads to other state boxes, decision boxes or conditional output boxes.

The DECISION BOX describes inputs to the state machine. The structure of a decision box is given in Figure 3.10. Each decision box has two exit paths. One path is taken when the enclosed condition is true and the other when the condition is false. These two paths are usually indicated by 1 for true and 0 for false. The inputs are also called qualifiers in the sense that they qualify an output or a transition. It should be stressed at this point that the exit paths in no way describe time dependence. They only represent logic relationships. The state box is the only element representing time.

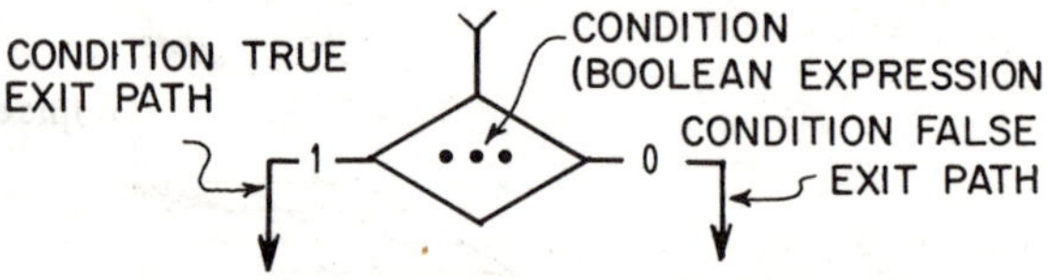

Figure 3.10 The Decision Box for a Single Decision.

The CONDITIONAL OUTPUT BOX describes other outputs which are dependent on one or more inputs in addition to the state of the machine. The structure of a conditional output box is given in Figure 3.11. The rounded corners of a conditional output box differentiate it from a state box. The output list in a conditional output box can also have immediate and delay operations.

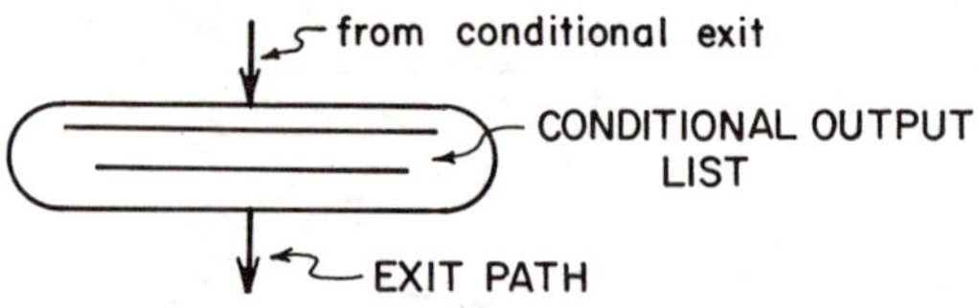

Figure 3.11 The Conditional Output Box.

An ASM BLOCK is a structure consisting of one state box and a network of decision boxes and conditional output boxes. An ASM block has one entrance and any number of exit paths represented by the structure of the decision boxes as shown in Figure 3.12. An ASM Chart consists of one or more interconnected ASM blocks. Each ASM block exit path must connect to a state. Each possible path from one state to the next is called a LINK PATH. There are one or more link paths per exit path depending on the decision box structure.

One ASM block describes the state machine operation during one state time. Each ASM block represents the present state, X, the state outputs, f[X], the conditional outputs, f[X,Q], and the next state, g[X,Q], for a set of inputs, Q, of the general state machine. Thus, the total f and g functions are described by the ASM chart on a state-by-state basis.

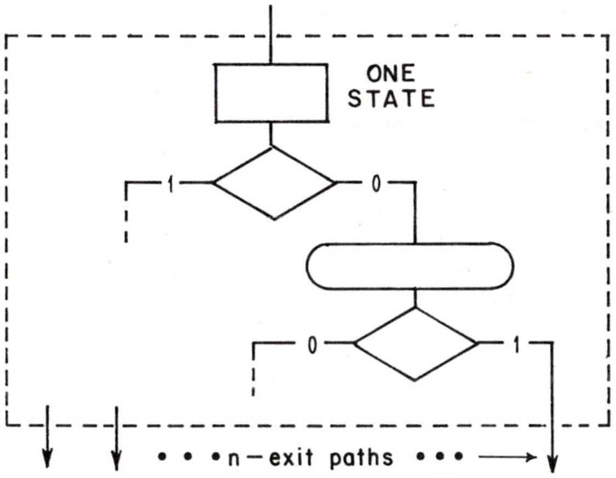

Figure 3.12 The ASM Block. Each ASM describes the state machine operation during one state time.

One almost obvious restriction is placed upon the interconnection of blocks in an ASM chart. This restriction is that there must be only one next state for each state and stable set of inputs. For example, Figure 3.13 shows a state transition that is undefined for a single state vector because two simultaneous next states are specified. A single machine cannot be in two states simultaneously as this chart would seem to imply. Similarly, if the decision boxes are structured so that two simultaneous next states are indicated for any single state and set of inputs, then this structure is also undefined. For example, Figure 3.14 shows an ASM block of a state, (x), and two inputs, YQ1 and YQ2, where two simultaneous next states are specified

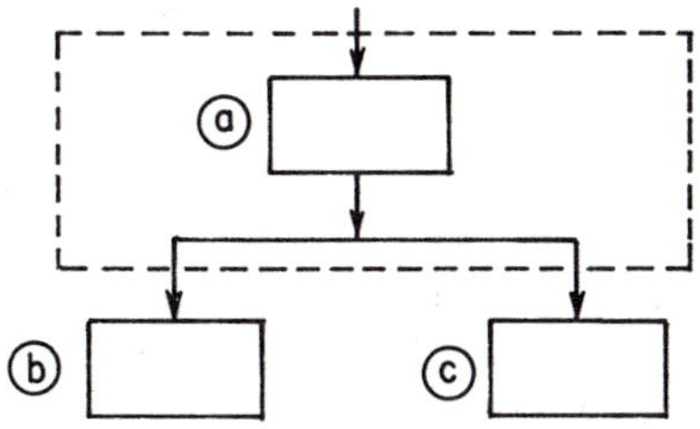

Figure 3.13 An Undefined State Transition.

when YQ2 = 1. Both states (y) and (z) are indicated as the next state regardless of the value of YQ1. As before, this structure is meaningless. The restriction of a single next state is really a logical extension of the desire to express an algorithm which has each step well defined. In this sense, two simultaneous activities are undefined when specified in a single algorithm.

Figure 3.15 shows two state descriptions which are similar to figure 3.14 but are each properly defined to allow only one possible next state.

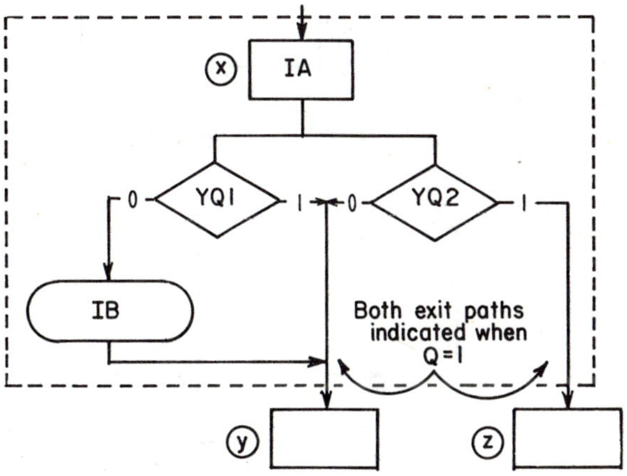

Figure 3.14 An Undefined Link-Path Condition.

Parallel and series connection of multiple qualifiers in an ASM block emerge as two equivalent forms of description. Figure 3.16 shows an example ASM block description using each form. Both descriptions produce the same machine operation. The choice of form is based on the ease of interpretation in a particular application. However, should there be any doubt concerning the existence of a single next state when using a parallel form, always revert to the series form to verify the description. A strict series form never leads to more than one next state for any condition of qualifiers.

Care should be taken to avoid some confusing structures. For example, Figure 3.17 describes an ASM block which has one condition box which loops on itself. Although the intent of this description is clear, that is, the transition to state (y) is made when YQ1 = 1, the literal description of the next state is confusing when YQ1 = 0. The condition box points to itself which is not a state. An equivalent and correct representation of the same state is given in Figure 3.18. This example points to one more restriction on an ASM chart, which can be stated as: every path depicted by a set of condition boxes must lead to a state. Otherwise the arrangement of condition boxes and conditional outputs in an ASM block is quite flexible.

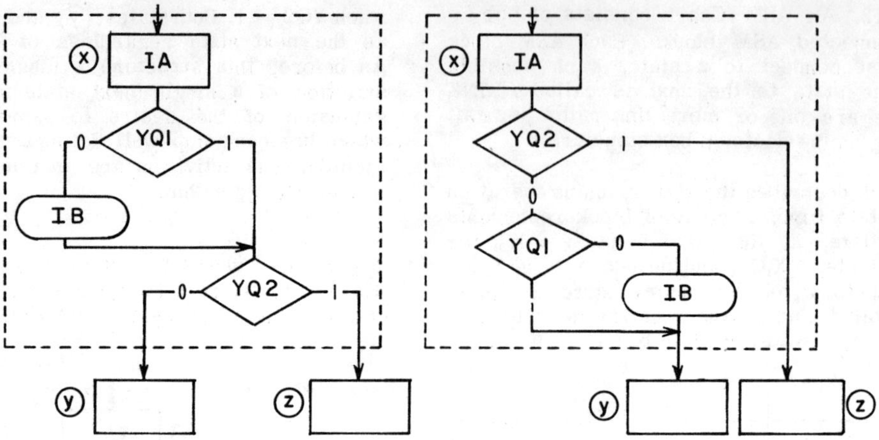

Figure 3.15 Two Possible Link-Path Structures.

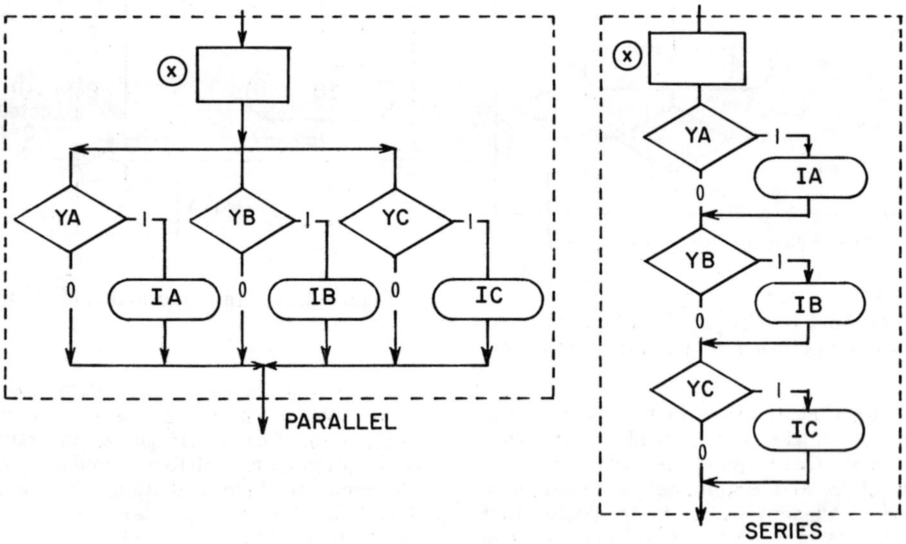

Figure 3.16 Two Equivalent Descriptions of an ASM Block.

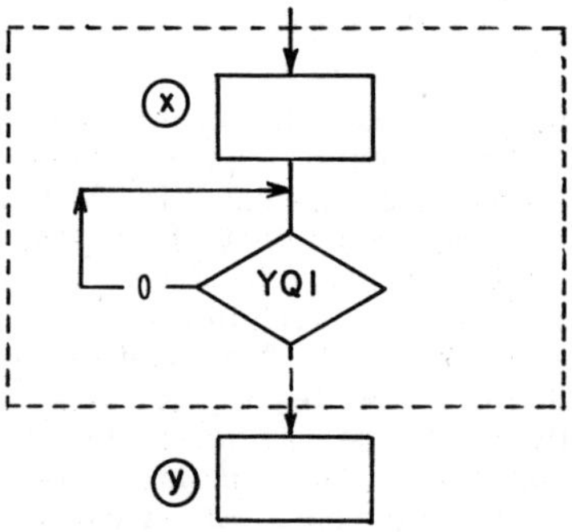

Figure 3.17 A Confusing Representation of the Next State.

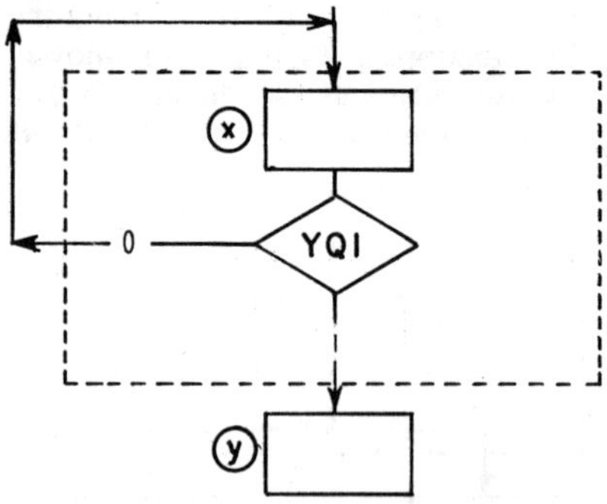

Figure 3.18 A Clear Representation of the Next State.

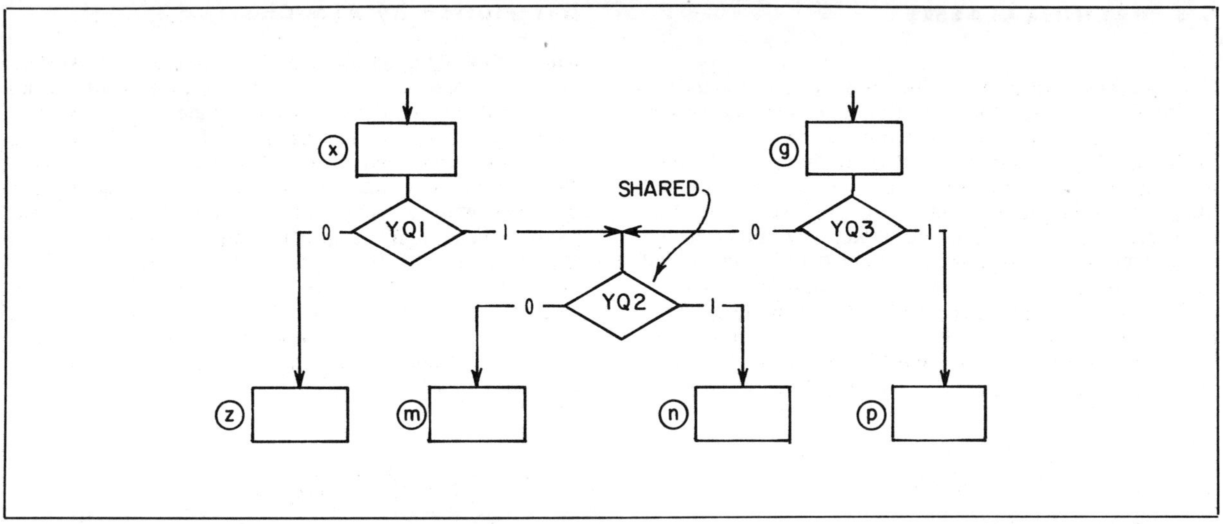

Figure 3.19 Shared Condition Boxes.

In actual practice, a couple of shortcuts are taken in drawing the ASM chart. First, the dotted lines outlining an ASM block are usually omitted because the block is clearly defined to be all the conditional boxes between one state and the next. Second, some of the condition boxes from one block may be shared by another as illustrated in Figure 3.19. This sharing does nothing to simplify the algorithm. It just saves writing.

The ASM chart fits in with the three classical description languages as shown in Figure 3.20. The chart shows the order of conversion ease which will be described in following sections. The ASM chart, the table, the map and the equation are each capable of describing algorithmic state machines. Even though the information described is the same in each of these languages, each emphasizes a particular aspect of the machine which is capitalized upon for synthesis in Chapters 4 and 5.

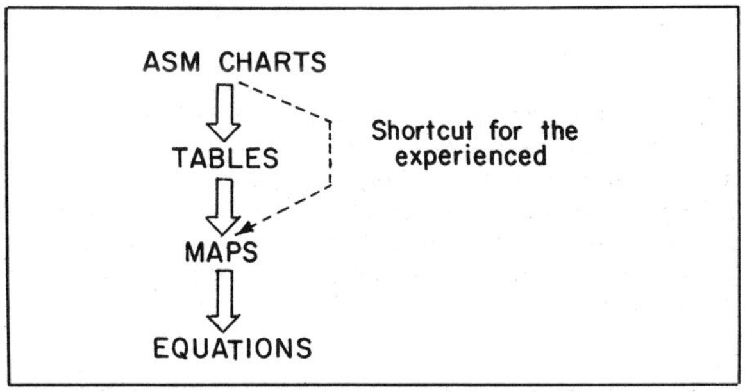

Figure 3.20 The Conversion Order for the Four Description Languages.

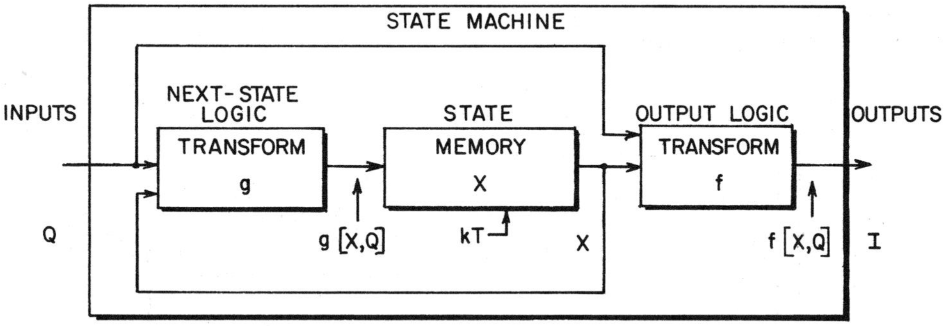

Figure 3.21 The General State Machine.

3.3 MACHINE CLASSES

The general state machine model, introduced in Chapter 2, is divided into five classes, which are desirable to unify the various types of logic and to organize the explanation of their operation. Each machine class is made from a portion of the general state machine model shown in Figure 3.21. For example, class 0 has no next-state function or state memory. The machine is thus fully described by the output logic alone. In a similar manner, additional elements are included for higher class machines until class 4 includes all the elements in the general model. The five classes are chosen to define particular types of machine behavior. Briefly, they are called:

- Class 0 - Combinatorial
- Class 1 - Delay
- Class 2 - Direct state transition and state output
- Class 3 - Conditional state transition and state output
- Class 4 - Conditional state transition and conditional state output

Any module can be identified as one of these classes and have connected with that identification a particular set of characteristics which proves valuable in relating one module to another. In the following sections, the description of each machine class is considered in detail using Boolean expressions, tables, maps and ASM charts.

3.4 CLASS 0 - COMBINATORIAL[3] LOGIC

Class 0 state machines have outputs which are a function of the inputs alone, I = f[Q]. The portion of the general state machine included in class 0 is shown by the solid lines in Figure 3.22.

[3] The words "combinatorial" and "combinational" are used interchangeably in logic design.

Description by ASM Chart

Any class 0 machine can be described by an ASM chart by one ASM block. The single state of this block has no particular significance other than representing that the next-state function is g[X,Q] = 1. The condition boxes describe the useful function. The usable outputs are always conditional outputs because with a single state any state output would never change. A simple example machine is given in Figure 3.23. In Figure 3.24 the single decision box contains the complete function for this machine, while Figure 3.25 is an equivalent representation in which each input variable is given a separate decision box. In either representation, IH is given when B + (E · C) is true.

Figure 3.23 An Example Class 0 Machine.

Although the ASM chart description of a Boolean function may be unconventional, its pictorial usefulness will be valuable as the more complex machines are discussed.

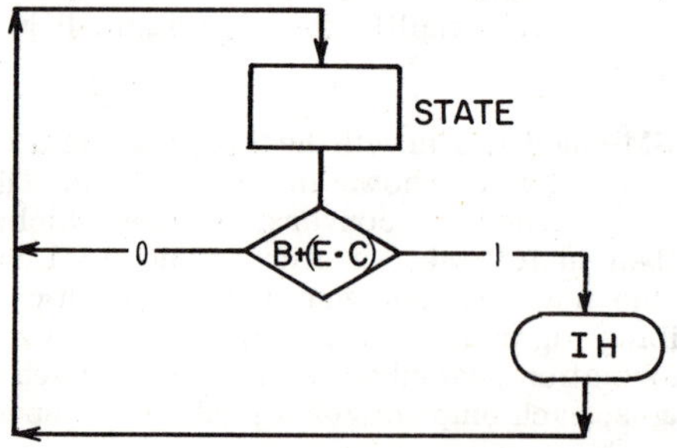

Figure 3.24 A Single Decision Box Description.

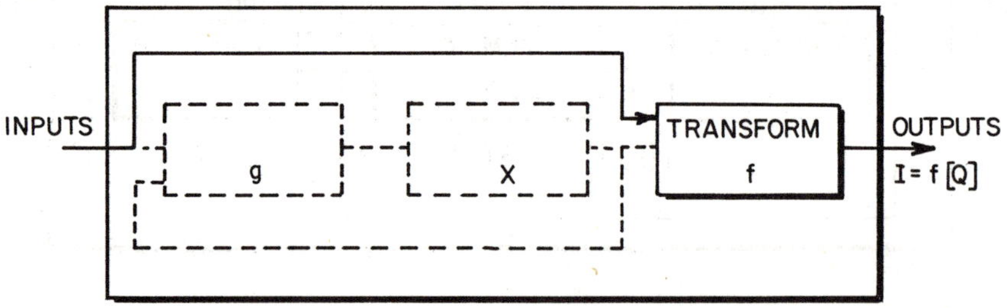

Figure 3.22 The State Machine Model for Class 0.

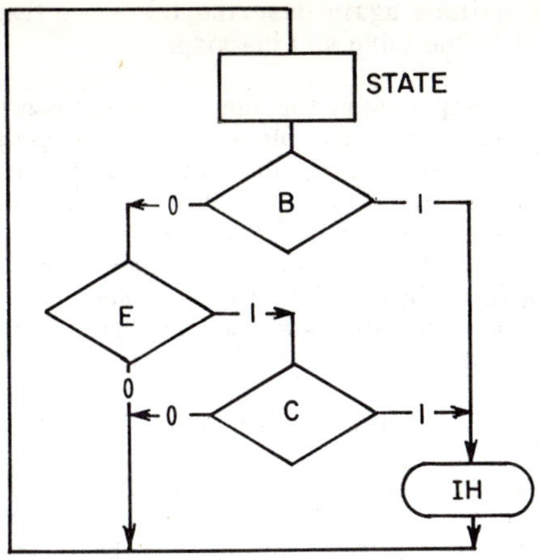

Figure 3.25 A Multiple Decision Box Description of Figure 3.24.

INPUT = $B \frown E \frown C$	OUTPUT
000	0
001	0
010	0
011	1
100	1
101	1
110	1
111	1

Figure 3.26 A Simple Tabular Description.

Description by Table

The ASM chart of a class 0 machine may be converted to a table by listing the conditional outputs resulting from all the possible input combinations. Figure 3.26 gives such a table for the simple example described previously. This table can be generated by tracing the ASM chart link paths for each combination of inputs.

Sometimes the function description is unknown. In these situations, the table can still be used to describe the function. For example, in Figure 3.27 a multiple output function is described to generate 10 different seven-segment characters for a display device where each output, a through g, corresponds to one of seven lights, and an output of 1 indicates that the segment indicated on the right of Figure 3.27 is to be lit. The outputs are not designated for the inputs which are never to occur. These are unused input combinations. The table adequately describes the character formation even though no other description may be known.

Q				f[Q]							Number Displayed
D	C	B	A	a	b	c	d	e	f	g	
0	0	0	0	1	1	1	1	1	1	0	0
0	0	0	1	0	1	1	0	0	0	0	1
0	0	1	0	1	1	0	1	1	0	1	2
0	0	1	1	1	1	1	1	0	0	1	3
0	1	0	0	0	1	1	0	0	1	1	4
0	1	0	1	1	0	1	1	0	1	1	5
0	1	1	0	1	0	1	1	1	1	1	6
0	1	1	1	1	1	1	0	0	0	0	7
1	0	0	0	1	1	1	1	1	1	1	8
1	0	0	1	1	1	1	0	0	1	1	9
1	0	1	0	-	-	-	-	-	-	-	
1	0	1	1	-	-	-	-	-	-	-	
1	1	0	0	-	-	-	-	-	-	-	
1	1	0	1	-	-	-	-	-	-	-	
1	1	1	0	-	-	-	-	-	-	-	
1	1	1	1	-	-	-	-	-	-	-	

← UNUSED INPUT COMBINATIONS

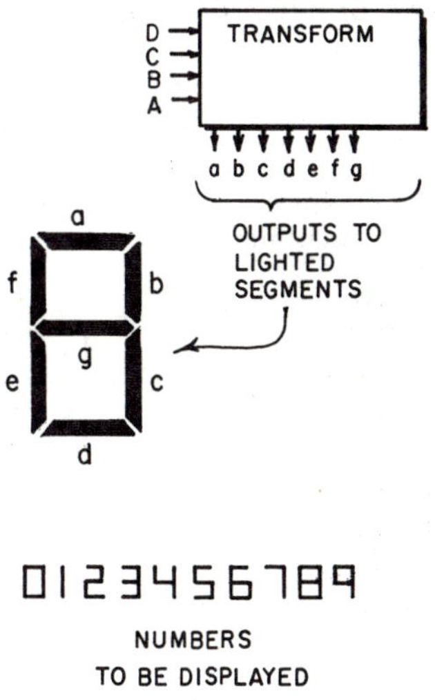

Figure 3.27 A Function to Generate Seven-Segment Characters.

Description by Maps

In Chapter 4 the map description is valuable for finding equations for transform operations. The map is valuable in going between a table and equations. The seven outputs from Figure 3.27 can be described by seven maps, one for each output, as shown in Figure 3.28. The maps are formed by entering the output corresponding to each input combination. Each square of the map corresponds to one input combination. These maps are another way to describe the seven-segment transform.

Description by Equation

There is a set of methods for going from the map to the equation which are described in Chapter 4. These methods yield one equation for each map description. In the seven-segment display transform, seven equations can describe the outputs in terms of the inputs. These equations are:

$$a = D + B + (\overline{A} \cdot \overline{C}) + (C \cdot A)$$

$$b = \overline{C} + (\overline{A} \cdot \overline{B}) + (A \cdot B)$$

$$c = A + \overline{B} + C$$

$$d = (\overline{A} \cdot \overline{C}) + (C \cdot A \cdot \overline{B}) + (\overline{C} \cdot B) + (\overline{A} \cdot B)$$

$$e = \overline{A} \cdot (B + \overline{C})$$

$$f = D + (\overline{A} \cdot \overline{B}) + (C \cdot \overline{B}) + (C \cdot \overline{A})$$

$$g = D + (\overline{A} \cdot B) + (C \cdot \overline{B}) + (B \cdot \overline{C})$$

These equations again describe the same transform specified by the table and the maps.

A Boolean expression for the link paths can be read directly from any ASM block by forming the AND of all the decisions along the length of the link path. For example, in Figure 3.25 there are four possible link paths as shown in Figure 3.29. Link paths (1) and (2) produce an output IH. Since these are the only possible link paths which produce this output, the equation for IH must be the logical OR of the two link-path expressions. The sum of the decisions along link path (1) is B and along link path (2) is B ·E ·C. An equation for IH, then, must be the OR (logical sum) of these two expressions, which is $IH = B + (\overline{B} \cdot E \cdot C)$. This equation is equivalent to $IH = B + (E \cdot C)$, as will be proven in Chapter 4. Similarly, an equation for $\overline{IH}$ can be produced by the OR of link paths (3) and (4), which is $\overline{IH} = (\overline{B} \cdot \overline{E}) + (\overline{B} \cdot E \cdot \overline{C})$.

3.5 CLASS 1 - DELAY MACHINES

The delay, Class 1, machine is discussed here as a combinatorial circuit with added state time delay. The PARTIAL NEXT-STATE EQUATION is used to describe the next-state behavior. The concept of synchronizing is covered along with the concepts of SYNCHRONOUS and ASYNCHRONOUS machines.

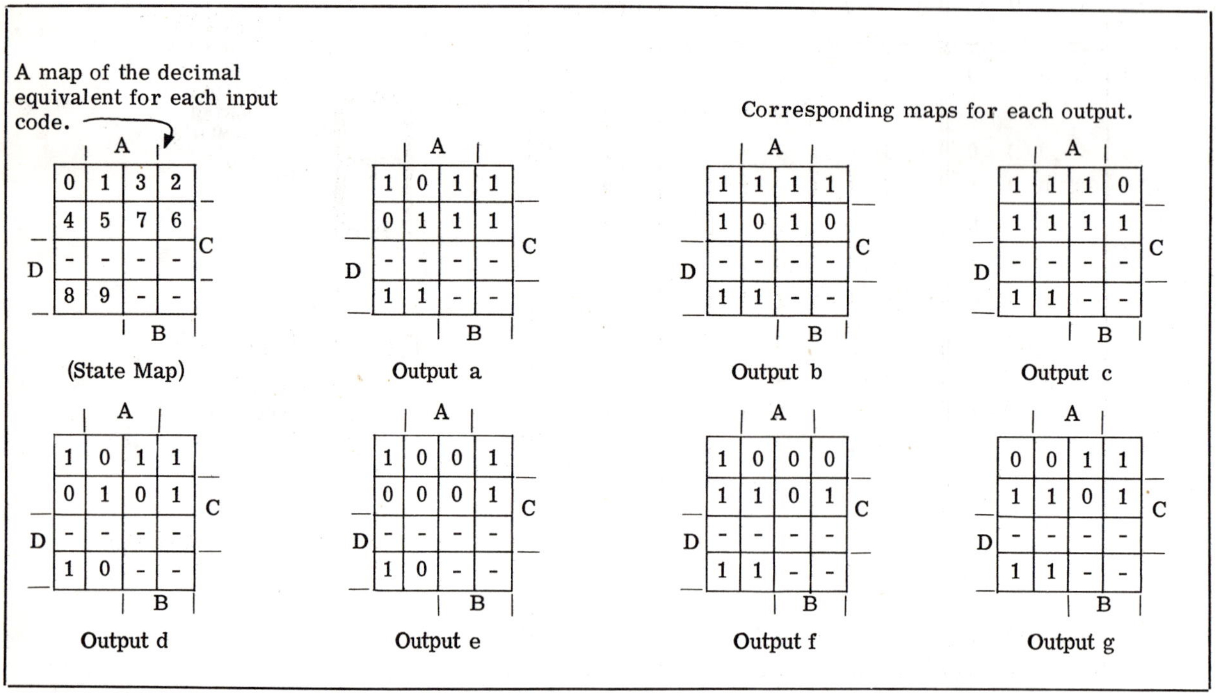

Figure 3.28 A Map Description of a Seven-Segment Character Generator also Described in Figure 3.26.

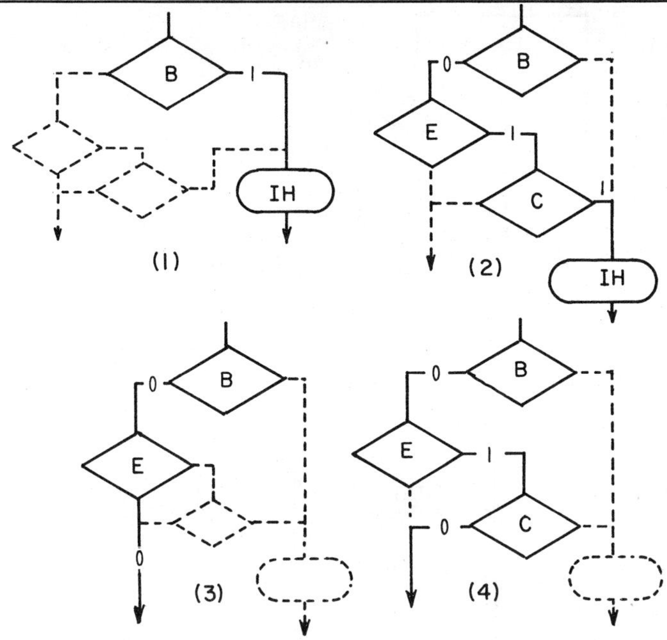

LINK PATH

(1) = IH = B

(2) = IH = $\overline{B} \cdot E \cdot C$

$\therefore$ IH = B + ($\overline{B} \cdot E \cdot C$)

LINK PATH

(3) = $\overline{IH} = \overline{B} \cdot \overline{E}$

(4) = $\overline{IH} = \overline{B} \cdot E \cdot \overline{C}$

$\therefore \overline{IH} = (\overline{B} \cdot \overline{E}) + (\overline{B} \cdot E \cdot \overline{C})$

Figure 3.29 Obtaining a Boolean Equation from the Link Paths of an ASM Block.

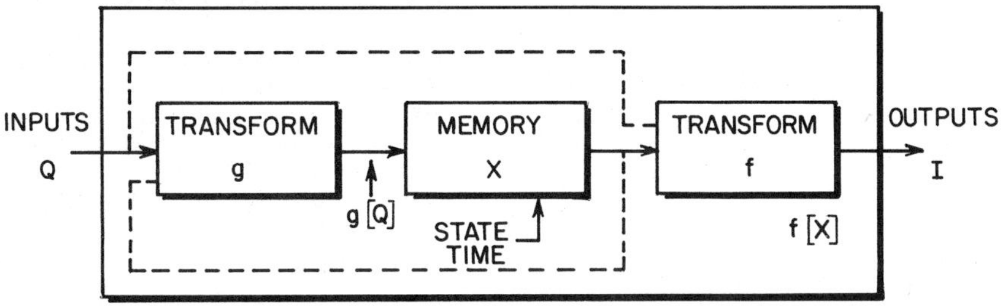

Figure 3.30 The General Class 1 Machine.

A Simplified Representation

Figure 3.30 describes the portion of a general state machine which is called a class 1 machine. It consists of one memory and two transform elements. Because the machine has no internal feedback, an equation can be written to describe the behavior involving only the inputs and the outputs.

This equation is I ← fg[Q] and does not involve the internal state, X. Therefore, the class 1 machine can be thought of as a combinatorial function, fg, which produces an output delayed by one state time, as compared to a class 0 machine which produces an output in the present state time. If X' represents the state of the combined function machine, then Figure 3.31 describes two ways to visualize the equivalent of the class 1 machine, either as a transform followed by a state delay on the outputs or as a state delay on the inputs followed by the transform. These representations are an aid to visualizing a delay machine function.

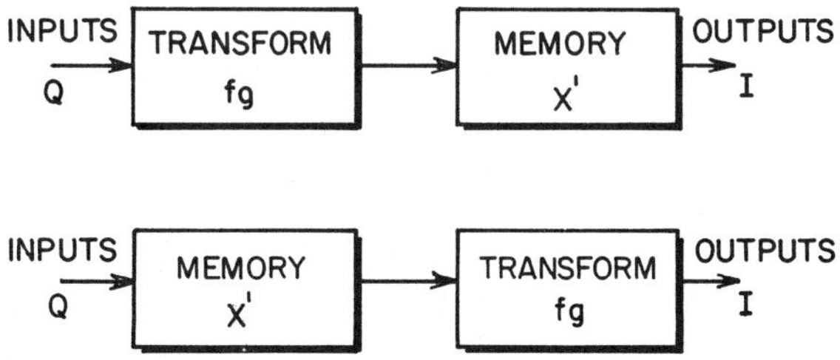

Figure 3.31 Two Representations for a Class 1 Machine Described by I←fg[Q] Which Are Functionally Equivalent to the Class 1 Machine in Figure 3.30.

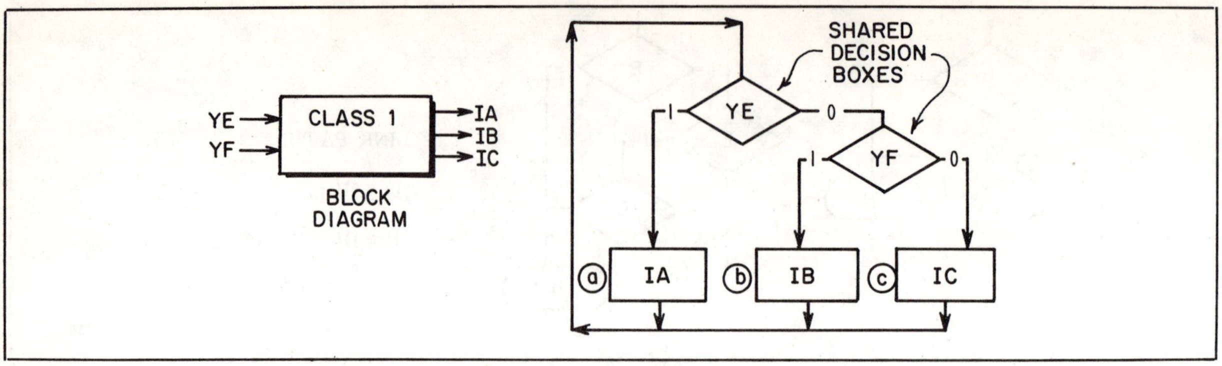

Figure 3.32 An ASM Chart Description of a Class 1 Machine.

Description by ASM Chart

In the ASM chart representation of the class 1 machine, the two functions f and g are described to show that the next state depends on the inputs alone and the output depends on the state alone. There is one ASM block for each state of X. The transform of the inputs to the next state is described by the condition box structure. Since there is only one next-state function and it depends on the inputs alone, the condition box structure is shared by every state. The outputs are indicated by the output list in each state. The ASM chart is rarely used to describe a class 1 machine.

A simple three-state class 1 machine is described by Figure 3.32. The shared decision box structure is shown above the states to emphasize the next-state behavior. Three outputs are generated; one for each state.

Description by Table

The delay machine table description is identical to the description of the combinatorial (class 0) machine except that the outputs occur in the next state. For example, Figure 3.33 describes a delay function having one output, IH, which may be compared to the combinatorial description in Figure 3.26. Figure 3.34 describes the three state delay function from Figure 3.32.

INPUT B⌒E⌒C	NEXT-STATE OUTPUT, IH
000	0
001	0
010	0
011	1
100	1
101	1
110	1
111	1

Figure 3.33 A Simple Delay Machine.

INPUT YE⌒YF	NEXT-STATE OUTPUT IA⌒IB⌒IC
0 0	001
0 1	010
1 0	100
1 1	100

Figure 3.34 A Tabular Description of Figure 3.32.

Description by Map

The delay machine may also be described by a group of maps which relate the present set of inputs to the next-state output. A map of the three state delay machine described by the table in Figure 3.34 may be described by the maps in Figure 3.35. One map is drawn for each output.

IA output

		YE
	0	1
YF	0	1

IB output

		YE
	0	0
YF	1	0

IC output

		YE
	1	0
YF	0	0

Figure 3.35 A Map Description of Figure 3.34.

Description by Equation

The delay operator is used in equations to denote outputs occurring in the next state. The equations for the three delay outputs described by the maps in Figure 3.35 may also be described by three delay functions as follows:

$$YE \rightarrow IA$$

$$YF \cdot \overline{YE} \rightarrow IB$$

$$\overline{YF} \cdot \overline{YE} \rightarrow IC$$

The equation provides a convenient means for describing the most common of the delay machines, the register. The register is a collection of single delay elements that store an input vector. In a simple register, the output is the same as the input delayed by one state time. If Q represents the input vector (a parallel set of inputs) and I represents the output vector (a parallel set of outputs), then a register holding Q can be described by $Q \rightarrow I$. Usually additional description is required to include the effects of other control inputs which may reset the register, $00 \ldots 0 \rightarrow I$ set the register to 1's, $111 \cdots 1 \rightarrow I$, or complement the register (invert all 1's and 0's).

Using the register as a simple example, the terminology for the PARTIAL NEXT-STATE EQUATION can be described. The partial next-state equation is a notation used for describing the next-state behavior of single state variables in a more complex machine. In this notation, each state variable is given a name (letter or number) and this name is used as a subscript to identify that particular portion of the next-state function. The complete next-state function, g[X,Q], is described by the concatenation of the partial next-state functions. In a similar manner the complete output function, f[X,Q], can be described by the concatenation of the PARTIAL OUTPUT EQUATIONS. For example, if a five-variable register has STATE = $R5 \frown R4 \frown R3 \frown R2 \frown R1$, INPUT = $Q5 \frown Q4 \frown Q3 \frown Q2 \frown Q1$ and OUTPUT = $I5 \frown I4 \frown I3 \frown I2 \frown I1$, the description of the delay machine would consist of the next-state description and the output description as follows:

$$g[Q] = \begin{bmatrix} g_{R1}[Q] = Q1 \\ g_{R2}[Q] = Q2 \\ g_{R3}[Q] = Q3 \\ g_{R4}[Q] = Q4 \\ g_{R5}[Q] = Q5 \end{bmatrix}, \quad f[X] = \begin{bmatrix} f_{I5}[X] = R5 \\ f_{I4}[X] = R4 \\ f_{I3}[X] = R3 \\ f_{I2}[X] = R2 \\ f_{I1}[X] = R1 \end{bmatrix}$$

These expressions can be simplified to

$$g[Q] = \begin{bmatrix} R1 \leftarrow Q1 \\ R2 \leftarrow Q2 \\ R3 \leftarrow Q3 \\ R4 \leftarrow Q4 \\ R5 \leftarrow Q5 \end{bmatrix} \qquad f[X] = \begin{bmatrix} I5 = R5 \\ I4 = R4 \\ I3 = R3 \\ I2 = R2 \\ I1 = R1 \end{bmatrix}$$

which is the same as the vector forms,

$$g[Q] = [X \leftarrow Q] \quad \text{and} \quad f[X] = [I = X]$$

In these descriptions, the brackets are used to indicate a concatenation operation for compactness, rather than using the $\frown$ defined in Section 2.1.

An example of the use of partial next-state equations is made by describing the next-state behavior of particular state variables assigned to the three state delay machine shown in the first part of this section (Figure 3.32). The description of this machine is duplicated in Figure 3.36, which includes the state codes for the state variables G and H.

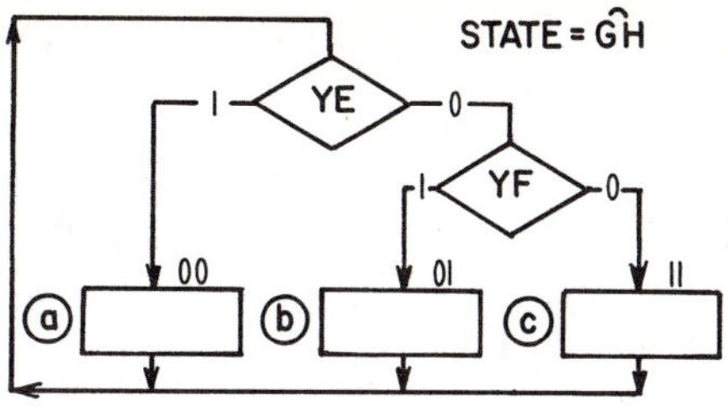

Figure 3.36 A Three State Delay Machine Including State Codes.

Figure 3.37 gives a tabular description of the machine including the state codes. The next state can be described with two maps, one for each variable as shown in Figure 3.38. Accordingly, the outputs can be described by maps in terms of the state variables as shown in Figure 3.39. These maps are partial functions as indicated.

INPUTS $YE \frown YF$		NEXT STATE $G \frown H$		NEXT OUTPUT $IA \frown IB \frown IC$		
0	0	1	1	0	0	1
0	1	0	1	0	1	0
1	0	0	0	1	0	0
1	1	0	0	1	0	0

Figure 3.37 A Tabular Description Including the State Codes.

$g_G[Q]$

		YE
	1	0
YF	0	0

$g_H[Q]$

		YE
	1	0
YF	1	0

Figure 3.38 The Partial Next States Described in Terms of the Inputs.

$I_{IA}[X]$

		G
	1	-
H	0	0

$I_{IB}[X]$

		G
	0	-
H	1	0

$I_{IC}[X]$

		G
	0	-
H	0	1

Figure 3.39 The Partial Outputs Described in Terms of the State Variables.

Again, each map is equivalent to equations as follows:

$$g[Q] = \begin{bmatrix} g_G[Q] = \overline{YE} \cdot \overline{YF} \\ g_H[Q] = \overline{YE} \end{bmatrix} \qquad I[X] = \begin{bmatrix} I_{IA}[X] = \overline{H} \\ I_{IB}[X] = H \cdot \overline{G} \\ I_{IC}[X] = G \end{bmatrix}$$

$$g[Q] = \begin{bmatrix} G \leftarrow \overline{YE} . \overline{YF} \\ H \leftarrow \overline{YE} \end{bmatrix} \qquad I[X] = \begin{bmatrix} IA = \overline{H} \\ IB = H \cdot \overline{G} \\ IC = G \end{bmatrix}$$

Therefore, the single description given by Figure 3.35 has been divided into a next-state function and an output function using the state variables in between. However, the two-part description still describes the same behavior between the inputs and the outputs. In the next sections, the two-part description will be the only way to describe the machine behavior since the internal state will affect the next-state behavior.

Before higher class machines are considered, the meaning of state time will be further explored.

Implications of a State Time

The delay and all higher class machines are characterized by an ability to hold the information existing during a short stable period for the entire state time following. For example, Figure 3.40 gives the time-line representation of the logic on two terminals of a delay machine described by YIN→ HOUT. The darkened regions before each state time indicate the required stable period[4] in the state time. Usually the stable period may represent a larger portion of the state time than shown. Whenever an input such as (a) in Figure 3.40 is a 1 in the stable region, the next-state output is a 1 as indicated by the arrow (b). Also, short pulses such as (c) are ignored when they occur in the transition period. The transition (d) occurs during the required stable period and this makes the next state undefined (either 1 or 0) until the next stable period, (e). Such a signal, which may be changing during the stable period, is called ASYNCHRONOUS in respect to the state-time system. The opposite definition is a SYNCHRONOUS signal. A machine which is based on an equal state time or a state time defined by external means (such as a clock) is called a SYNCHRONOUS machine. A machine whose state time depends solely upon the internal logic circuit delays is called an ASYNCHRONOUS machine.

3.6 CLASS 2 - DIRECT STATE TRANSITION AND STATE OUTPUT

A class 2 machine is a portion of a general state machine having its next state determined by the present state alone, $X \leftarrow g[X]$, and outputs determined by the present state alone, $I = f[X]$. The portion of the general state machine forming a class 2 machine is indicated by the solid lines in Figure 3.41. Synchronous counters are an example of class 2 machines. The important concept introduced in this section is the idea of a SEQUENCE of states.

[4] The required stable period is often called the setup time in flip-flops.

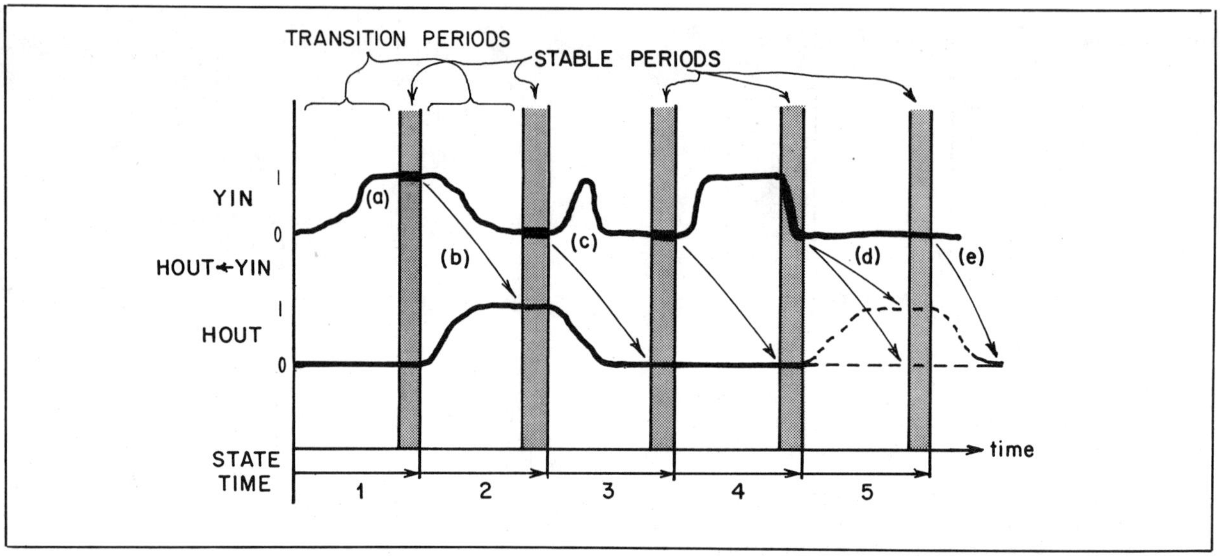

Figure 3.40 An Interpretation of the State Time as a "Snapshot" of Logic Behavior During Stable Periods.

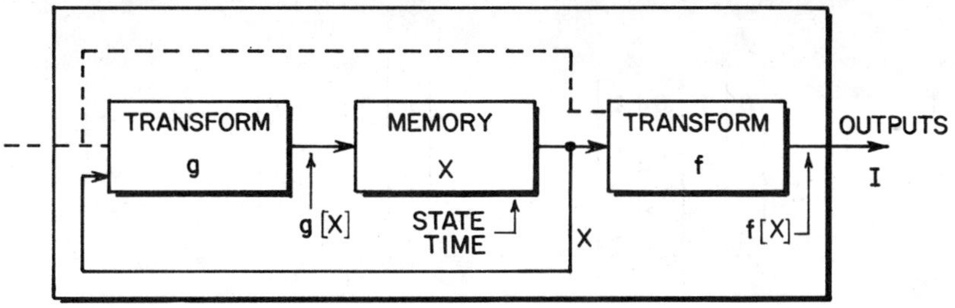

Figure 3.41 The State Machine Model for Class 2 Machines.

A Next-State Description Using an ASM Chart

Figure 3.42 shows an ASM chart description of the decade counter. It is easy to see that each state has only one next state and that the states are in a SEQUENCE. The state code is placed on the upper portion of the state box. A transition from one state to the next is made for each clock time which defines the state time. A decade counter counts in the sense that the number of clock inputs is remembered by the state. But, after 10 counts the states repeat. These repeating sequences of states are called CYCLES. The count of the counter can be determined by looking at the state variables which usually form the counter outputs. The absence of any condition boxes in the ASM chart results from the class 2 requirement that the next state be a function of the present state alone.

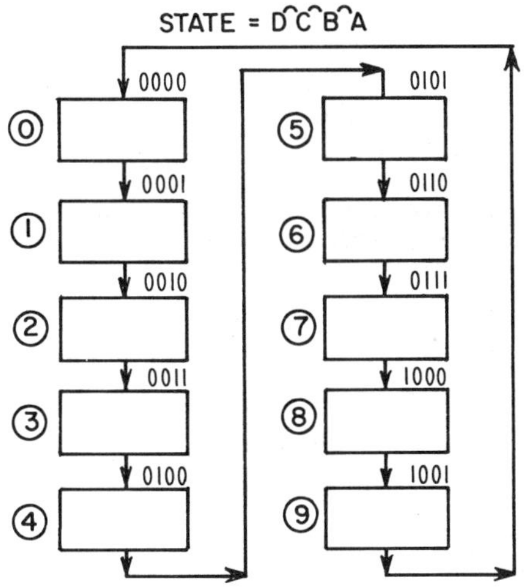

Figure 3.42 An ASM Chart Description of a Decade Counter.

A Tabular Description of the Next State

In forming a tabular description of the next-state function, it is essential to realize that in class 2 machines each present state has only one next state. Figure 3.43 gives a tabular description of the decade counter. For each present state, one next state is listed. In this counter, the outputs equal the present state; therefore, I = f[X] = X.

DECIMAL STATE TIME	PRESENT STATE X STATE VECTOR D⌒C⌒B⌒A				NEXT STATE g[X] STATE VECTOR D⌒C⌒B⌒A			
0	0	0	0	0	0	0	0	1
1	0	0	0	1	0	0	1	0
2	0	0	1	0	0	0	1	1
3	0	0	1	1	0	1	0	0
4	0	1	0	0	0	1	0	1
5	0	1	0	1	0	1	1	0
6	0	1	1	0	0	1	1	1
7	0	1	1	1	1	0	0	0
8	1	0	0	0	1	0	0	1
9	1	0	0	1	0	0	0	0

Figure 3.43 A Tabular Description of the Next-State Equation for a Decade Counter.

Next-State Description by Maps

Maps of the partial next state for each of the state variables may be made from the table in Figure 3.43. These four maps are shown in Figure 3.44.

A Next-State Description by Equation

A decade counter can also be described by a set of partial next-state equations. Although these equations are not readily evident at first glance, in Chapter 4 techniques will be described to obtain these equations from the map descriptions. The equations are:

$$g[X] = \begin{bmatrix} A \leftarrow \bar{A} \\ B \leftarrow (\bar{A}\cdot B)+(\bar{B}\cdot A\cdot\bar{D}) \\ C \leftarrow (\bar{B}\cdot C)+(\bar{A}\cdot C)+(\bar{C}\cdot A\cdot B) \\ D \leftarrow (\bar{A}\cdot D)+(A\cdot B\cdot C) \end{bmatrix}$$

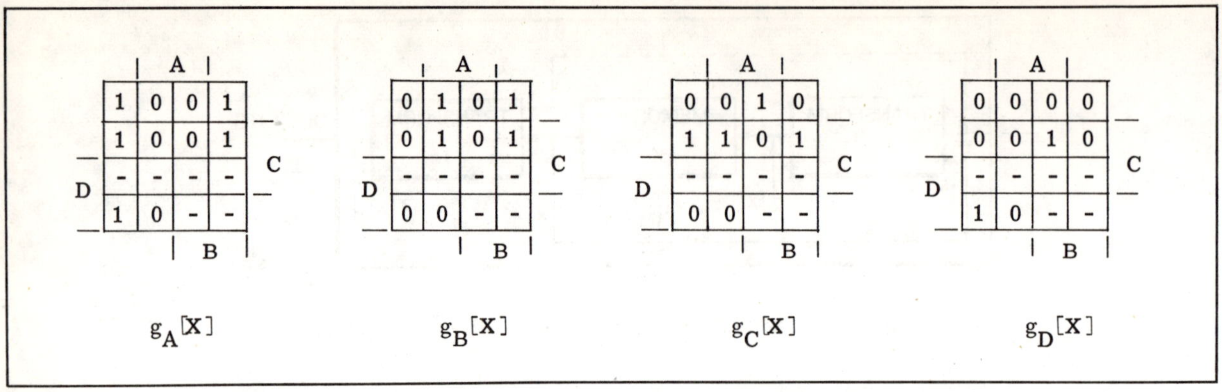

Figure 3.44 The Partial Next-State Maps of the Four State Variables of the Decade Counter from Figure 3.42 or 3.43.

Output Description

In a class 2 machine, state outputs (I = f[X]) can also be added. Figure 3.45 describes the decade counter with three added outputs: IHLOW for counts less than 5, IHHI for counts greater than or equal to 5 and IHZRO for zero counts. They are immediate instructions and are therefore preceded by an "I." The tabular description of the outputs is given in Figure 3.46. A map is made for each output by indicating in which states the output is desired, and the equations can be written as will be described in Chapter 4. These descriptions appear in Figure 3.47. All these descriptions are different ways of describing the same three outputs.

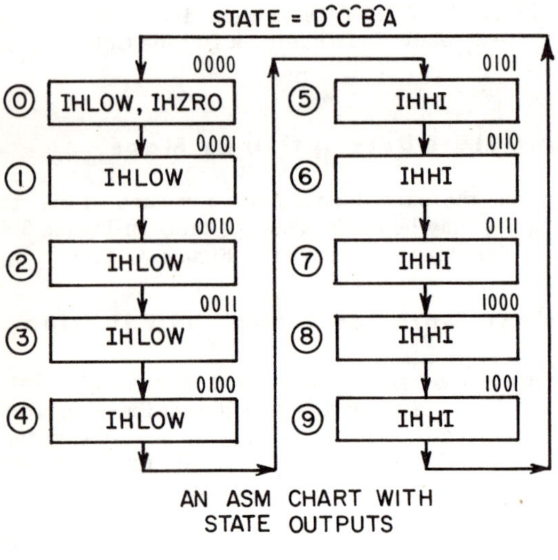

Figure 3.45 An ASM Chart Description of the Decade Counter with Three Outputs.

STATE	OUTPUTS		
D⌒C⌒B⌒A	IHZRO	IHLOW	IHHI
0 0 0 0	1	1	0
0 0 0 1	0	1	0
0 0 1 0	0	1	0
0 0 1 1	0	1	0
0 1 0 0	0	1	0
0 1 0 1	0	0	1
0 1 1 0	0	0	1
0 1 1 1	0	0	1
1 0 0 0	0	0	1
1 0 0 1	0	0	1

Figure 3.46 Tabular Description of State Outputs

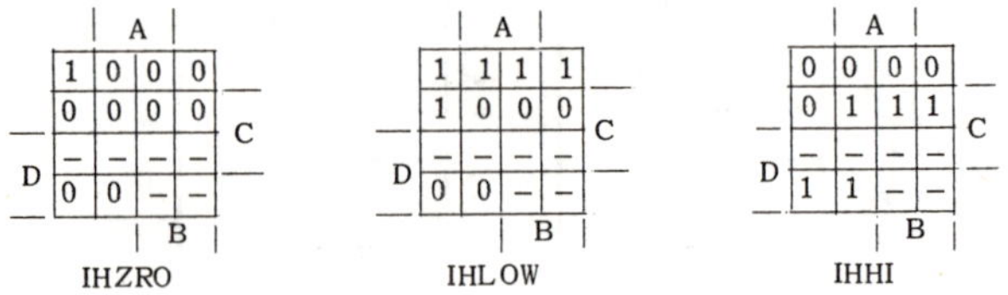

$IHZRO = \overline{A} \cdot \overline{B} \cdot \overline{C} \cdot \overline{D}$

$IHLOW = (\overline{C} \cdot \overline{D}) + (\overline{A} \cdot \overline{B} \cdot \overline{D})$

$IHHI = \overline{IHLOW}$

Figure 3.47 The Map and Equation Description of State Outputs.

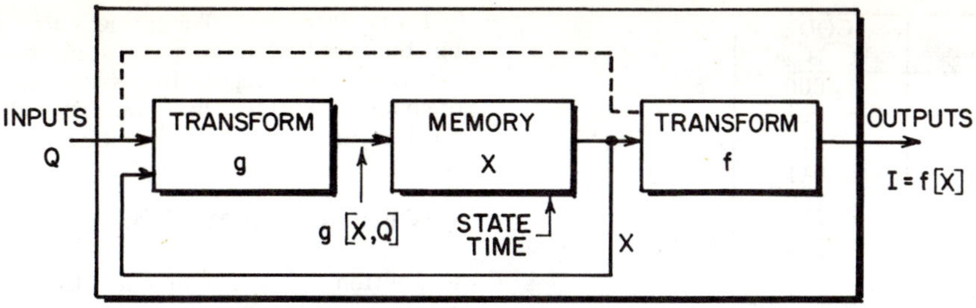

Figure 3.48 A State Machine Model for Class 3 Machines.

3.7 CLASS 3 - CONDITIONAL STATE TRANSITION AND STATE OUTPUT

Class 3 state machines have the next state determined by both the present state and the inputs, g[X,Q], and the outputs determined by the present state alone, f[X]. The class 3 state machine is indicated by the solid line portion of the general state machine in Figure 3.48. Class 3 machines are capable of choosing between alternative state sequences. They can perform all algorithms. Class 4 machines can only provide some simplifications.

A Description by ASM Charts and Tables

The conditional transition machine can be described by any of the four description languages already discussed (truth table, ASM chart, map or equation). This discussion starts with the ASM chart. Figure 3.49 is an example ASM chart of a simple class 3 machine which counts in a cycle of 5 or 8, depending on the input YC8. In a class 2 machine, there is no means for altering a sequence as done in this example.

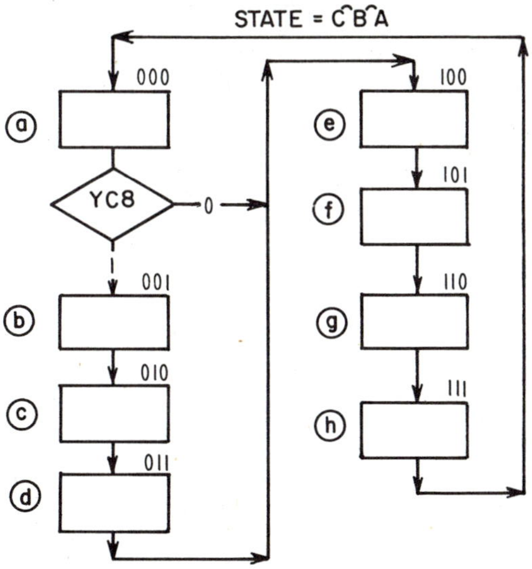

Figure 3.49 An ASM Chart of a Changeable Counter

There is a conditional transition from state ⓐ to ⓑ or ⓔ. All other transitions are direct. The truth table description in Figure 3.50 has an input column in addition to the present state and the next-state columns. Normally, there would be an output column as well, but in this example no outputs are described. Each state must be assigned a state code before a map description of the changeable counter can be made. This important step is called STATE ASSIGNMENT. Some assignments result in more efficient circuitry than others. The process of selecting efficient state assignments will be covered in Chapters 4 and 5.

One possible state assignment for the changeable counter is indicated by the placement of the state letters on a map of three state variables C, B, and A, as shown in Figure 3.51. This map is called a STATE MAP. The same information is given by a STATE TABLE, in Figure 3.52, and the ASM chart, in Figure 3.49. Using this state assignment, the next-state table is as shown in Figure 3.53.

INPUT, Q	STATE, X	NEXT STATE, g[X, Q]
YC8	a	b
$\overline{YC8}$	a	e
---	b	c
---	c	d
---	d	e
---	e	f
---	f	g
---	g	h
---	h	a

Figure 3.50 Next State for the Changeable Counter.

A (columns 2–3)

a	b	d	c	$\overline{C}$
e	f	h	g	C

B (columns 3–4)

Figure 3.51 The State Map.

STATE	CODE C B A
a	000
b	001
c	010
d	011
e	100
f	101
g	110
h	111

Figure 3.52 The State Table.

CURRENT STATE				NEXT STATE		
INPUT YC8	STATE C	B	A	g[X, Q] C	B	A
1	0	0	0	0	0	1
0	0	0	0	1	0	0
-	0	0	1	0	1	0
-	0	1	0	0	1	1
-	0	1	1	1	0	0
-	1	0	0	1	0	1
-	1	0	1	1	1	0
-	1	1	0	1	1	1
-	1	1	1	0	0	0

Figure 3.53 The Next-State Table Using the State Assignments.

3.7.2 A Description by Maps

The next-state equations may be described by using maps for the partial next states of each of the state variables. Therefore, three maps can describe the partial next-state equations of the changeable counter. These three maps are drawn in Figure 3.54. The partial next-state map for the state variable B is constructed exactly in the same way as described in Section 3.5. The next-state maps for the state variables A and C, however, are slightly different in that one of the map entries is the input YC8. The implication is that the next state of A and C depends on, or is conditional on, the input YC8. It is these two entries, $\overline{YC8}$ and YC8, that make this counter have an alterable sequence. If YC8 is a 0 and the counter is in state (a), then the next state of the A variable is a 0 and the next state of the C variable is 1. Since the next state of the B variable is always 0, the next state of the counter will be 100 or state (e). If YC8 had been a 1 in state (a), C would have a next state of 0, and A a next state of 1 making the next state

	A	A	
YC8	0	1	0
1	1	0	1
		B	B

C (bottom row)

NEXT STATE OF C
g_C[X,Q]

	A	A	
0	1	0	1
0	1	0	1
		B	B

C (bottom row)

NEXT STATE OF B
g_B[X,Q]

	A	A	
YC8	0	0	1
1	0	0	1
		B	B

C (bottom row)

NEXT STATE OF A
g_A[X,Q]

Figure 3.54 The Partial Next-State Maps for the Three State Variables of the Changeable Counter.

001 or state (b). Thus, the entry of input variables on the next state map is a method of describing in detail how the input affects the next state transitions of each variable. In Chapter 4, the entry of inputs on the map will be described as a method of SELDOMLY USED VARIABLES.

The Black Jack Machine Description

In Section 2.3 a set of operations was defined for a machine called the Black Jack Dealer, which can be described as a class 3 machine. The operations were divided into the immediate and the delayed types, as follows:

Immediate	Delayed
IHHIT	HTVC,HK50
IHSTND	HT10, HJ50
IHBRK	HT22
	HCLS

Using these operations and the qualifier defined in Section 2.3, an ASM chart can be constructed to describe the control algorithm as shown in Figure 3.55. The operation of the Black Jack control can be understood by tracing through the ASM chart using the definitions from Section 2.3 and the operational flow chart in Figure 2.6.

There are eight states in the ASM chart. The dotted lines indicate each ASM block although these lines are usually implied by the chart itself. State (d) has five exit and link paths (possible paths from one state to the next). Four other states have two link paths, and three states only have one link path. Thus, the total number of link paths is the sum of these links, which is 16.

A tabular description of the Black Jack control is given in Figure 3.56. This table was constructed by listing the next states and the conditions for each of the 16 link paths and the outputs for each state. Although both the table and the ASM chart contain the same information, it is usually easier to understand the operation of a machine from the ASM chart than from the table, while the table is a convenient first step in going to the circuitry which is discussed in Chapter 4.

3.8 CLASS 4 - CONDITIONAL STATE TRANSITION AND CONDITIONAL STATE OUTPUT

This section covers the last and most general type of machine, the class 4 machine, which has internal states with both the next state and the outputs determined by the inputs and the present state. This type of machine is the same as a general state machine which is described by a block diagram in Figure 3.57. Since the next-state function in this machine has the same form as a class 3 machine, the only part which has to be covered is the conditional state output.

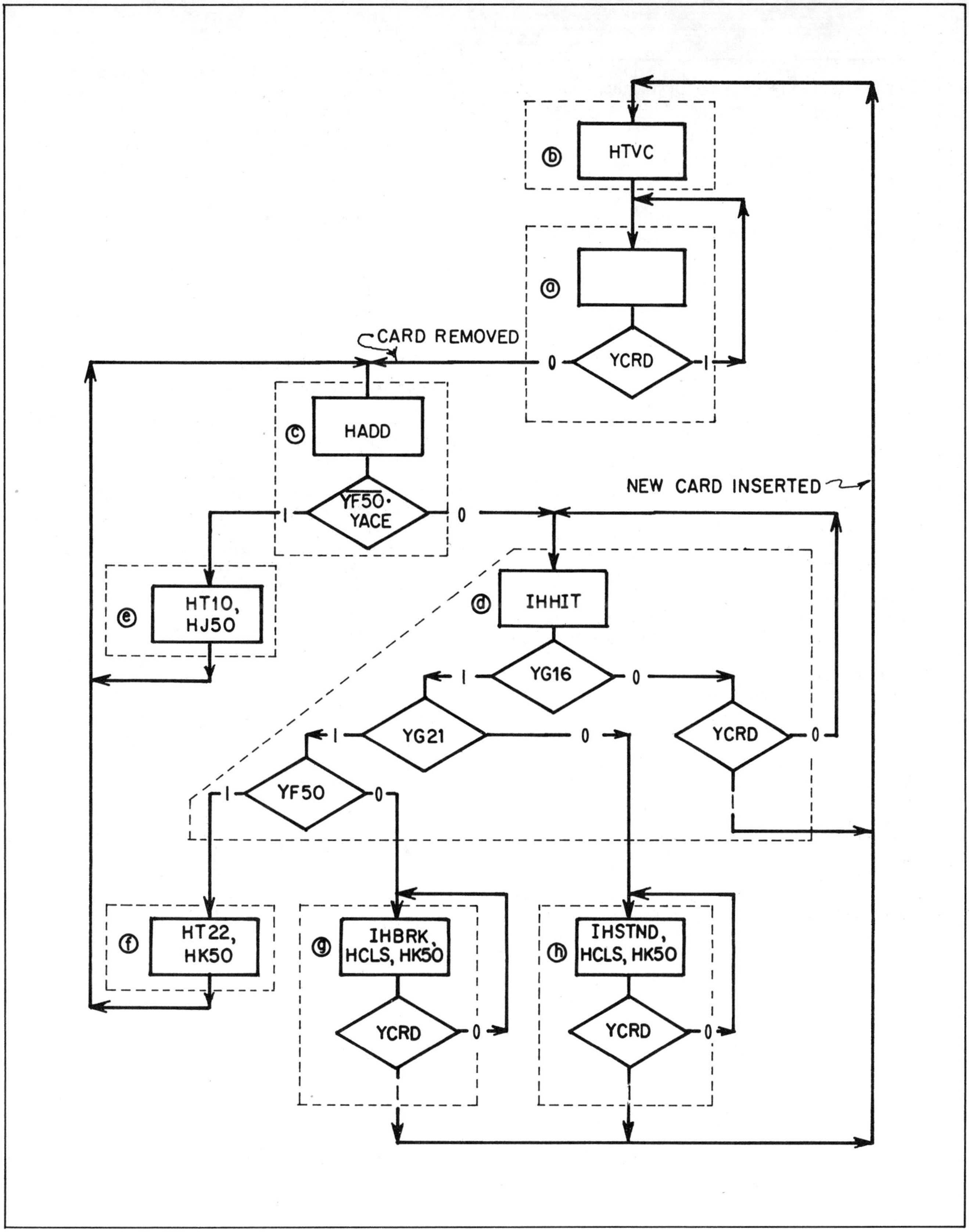

Figure 3.55 The Black Jack Control Algorithm Described as a State Output, Conditional Transition (Class 3) Machine.

X STATE	Q INPUTS YCRD	YF50	YACE	YG16	YG21	I INSTRUCTION OUTPUTS	g[X, Q] NEXT STATE
a	1	–	–	–	–	0	a
a	0	–	–	–	–	0	c
b	–	–	–	–	–	HTVC	a
c	–	0	1	–	–	HADD	e
c	– –	1 –	– 0	– –	– –	HADD	d
d	0	–	–	0	–	IHHIT	d
d	1	–	–	0	–	IHHIT	b
d	–	–	–	1	0	IHHIT	h
d	–	0	–	1	1	IHHIT	g
d	–	1	–	1	1	IHHIT	f
e	–	–	–	–	–	HT10, HJ50	c
f	–	–	–	–	–	HT22, HK50	c
g	0	–	–	–	–	IHBRK, HCLS, HK50	g
g	1	–	–	–	–	IHBRK, HCLS, HK50	b
h	0	–	–	–	–	IHSTD, HCLS, HK50	h
h	1	–	–	–	–	IHSTD, HCLS, HK50	b

Figure 3.56 Tabular Description of the Black Jack Machine from the ASM Chart in Figure 3.55.

The Use and Description of Conditional State Outputs

An ASM chart description of a class 4 machine is very similar to that of a class 3 machine with the addition of conditional outputs in the ASM block link structure. This means that one state may produce several different outputs as a function of the inputs, I = f[X,Q]; and this ability often enables an algorithm to be described with fewer states than with state outputs alone.

Figure 3.58 provides a comparison of the effects of a conditional and a state output. In Figure 3.58 (a), the instruction HINC is executed in state (x) regardless of the condition YTEN, while in Figure 3.58 (b), HINC is executed only when YTEN = 0. The difference in these two descriptions is that the counter is set to 11 (the counter counts higher than 10) when entering state (y) in Figure 3.58 (a) and is only set to 10 when entering state (y) in Figure 3.58 (b). It is clear that the conditional output allows modification of instructions on the basis of inputs. This ability greatly expands the flexibility of issuing instructions, which explains why a design can often be realized in fewer states when using conditional outputs.

The tabular description of a class 4 machine is changed very little from that of a class 3 machine. There is still one table row for each link path. The map descriptions follow from the table and the equations from the maps as before.

Conditional Outputs in the Black Jack Description

In the Black Jack control (Figure 3.55), states (e) and (f) can be made conditional outputs on states (c) and (d), respectively, as shown in Figure 3.59. This change eliminates two link paths and two states. The operation of the Black Jack control is the same. States (e) and (f) were not really required to separate any operations since HT10 and HADD and J50 may be performed simultaneously in state (c), as may IHHIT and IT22 and K50 in state (d). These operations may be represented as follows:

In state (c) when E50 · YACE = 1:

SUM ← SUM (+) CARD
CARD ← Decimal 10
F50 ← 1
STATE ← (c)

} simultaneously at the end of the state time

In state (d) when YG16 · YG21 · F50 = 1:

IHIT = 1 ⟶ During state (d)

CARD ← Decimal -10
F50 ← 0
STATE ← (c)

} simultaneously at the end of the state time

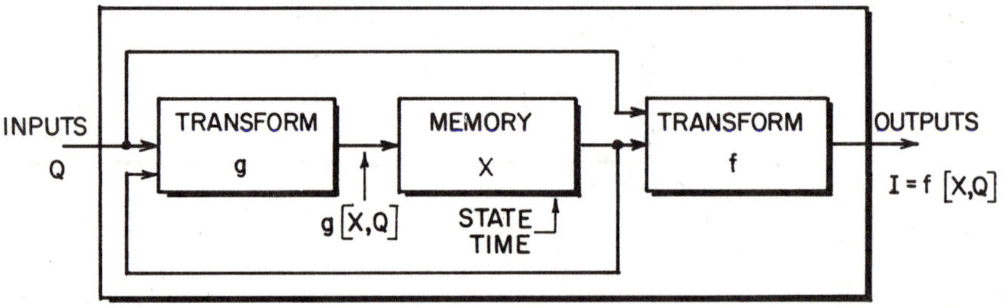

Figure 3.57 The General State Machine for Class 4 Machines.

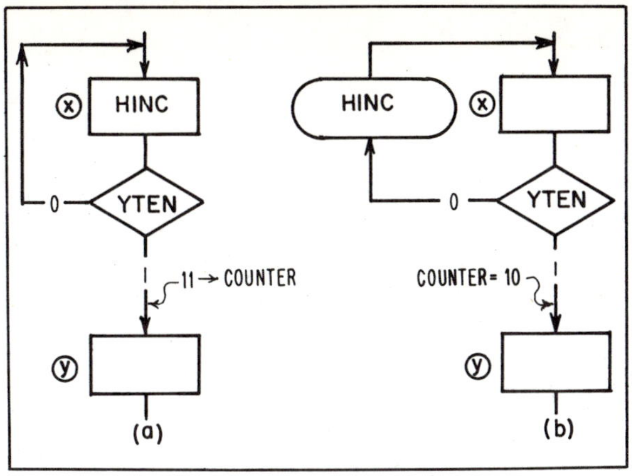

Figure 3.58 A Conditional Output Comparison.

3.9 SUMMARY OF MACHINE CLASSES

Four languages for the description of state machines have been introduced. The languages are the Boolean expression, the table, the map and the ASM chart. For each of the five classes of state machines, these languages were used to describe the machine's operations and state sequence. The ASM chart was particularly useful in the description of class 2, 3, and 4 machines.

The five machine classes were given a name corresponding to the equations representing the next-state function and the output function. These relationships are summarized by Figure 3.60.

CLASS	STATE MACHINE FUNCTIONS	NAME OF MACHINE CLASS
Class 0	I = f[Q] X ← 1	Combinatorial output
Class 1	I = f[X] X ← g[Q]	Delay
Class 2	I = f[X] X ← g[X]	State output, direct state transition
Class 3	I = f[X] X ← g[X,Q]	State output, conditional state transition
Class 4	I = f[X,Q] X ← g[X.Q]	Conditional state output, conditional state transition

Figure 3.60 A Summary of Machine Classes.

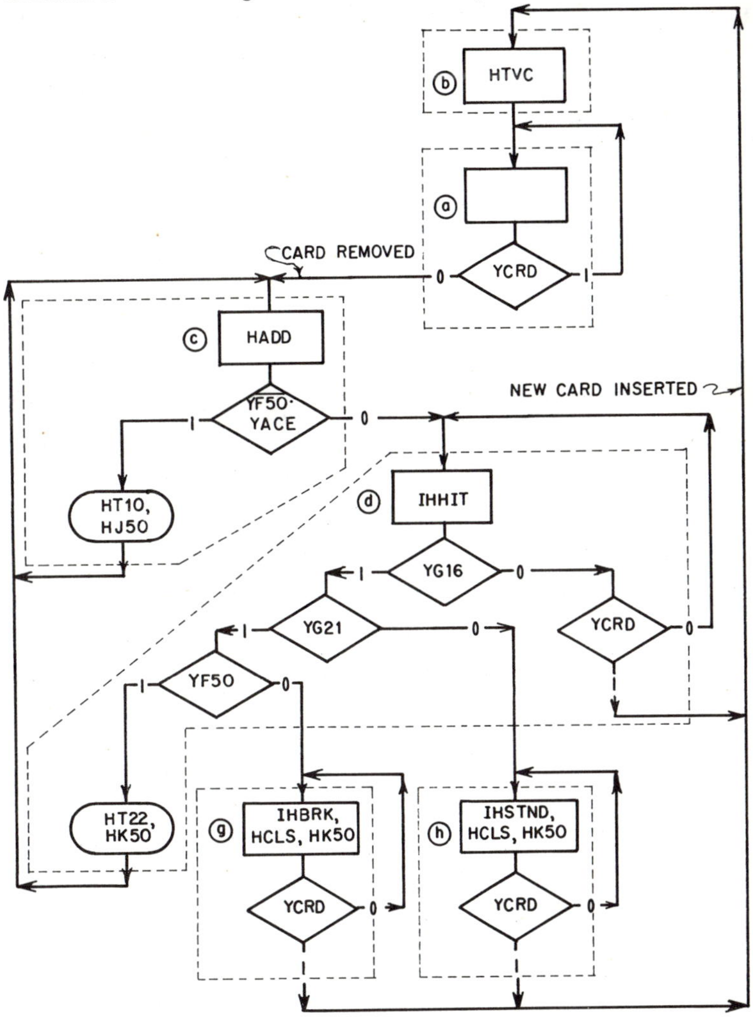

Figure 3.59 The Black Jack Machine Described Using Conditional Outputs to Replace States ⓔ and ⓕ in Figure 3.55.

CHAPTER IV
SYNTHESIS FOR GATE-ORIENTED DESIGN

4.0 THE FOURTH PHASE

After definition, description and evaluation comes synthesis, the fourth phase of logic design. Synthesis is a broad study of translating carefully prepared logic descriptions into physical machines to perform the desired LOGIC FUNCTIONS. In this chapter the basic element in the realization is the GATE. A collection of gates is represented by a group of symbols making up a SYMBOLIC LOGIC CIRCUIT. A set of translations is described for converting descriptions into logic circuits. The basic areas covered are equation-to-gate conversions, map simplifications, output function synthesis, next-state function synthesis, flip-flops, STATE ASSIGNMENT and HAZARDS.

4.1 SYMBOLIC LOGIC CIRCUITS

A set of symbols is given in this section for describing circuits to perform logic. The symbols are interconnected by lines indicating wires at one of two potentials. These wires are called LOGIC LINES. A group of interconnected symbols is called a LOGIC CIRCUIT.

Specific symbols represent circuits that perform a given Boolean function as determined from a given definition of the logic levels for the circuits inputs and outputs. The standard used here is that the more positive level is the logical 1 level for all logic lines.[1]

Figure 4.1 shows an electrical circuit to illustrate the process. The circuit need not be understood but is given for those who might be interested. The table below the circuit describes the behavior of the input and the output logic lines as determined from the definition for the 0 and 1 given above it. The Boolean equation for this table is $C = \overline{A \cdot B}$. This circuit is known as a NAND (NOT AND) circuit. Three other basic types of circuits used in logic are called the AND, the OR and the NOR (NOT OR). Each of these circuit types is called a GATE.

The symbols for the AND, NAND, NOR and OR circuits are given in Figure 4.2. Each symbol has a characteristic shape describing its logic function. Information may be written into the symbol to reference different circuit implementations of the same gate function or to refer to the physical location of the circuit. This designation will be called the TAG.

[1] When the more positive level is used as logic 1, it is called "positive logic", and when the more negative level is used as logic 1 it is called "negative logic". A consistent use of either convention will yield equivalent logic. This book uses positive logic throughout.

The tag is defined for each set of logic circuits. Many circuits used for logic functions do not fall into one of the four types mentioned. These circuits are referenced by a rectangular symbol and a tag, as also shown in Figure 4.2. The inputs are denoted by small arrows, and the outputs are unmarked except where reference to some special explanatory note is required. The logic function of these circuits is explained by Boolean equations, tables, maps, ASM charts or other logic circuits. In this manner, very complex electrical circuits may be reduced to a much simpler representation as a logic circuit while still maintaining adequate references to suitable logic function descriptions.

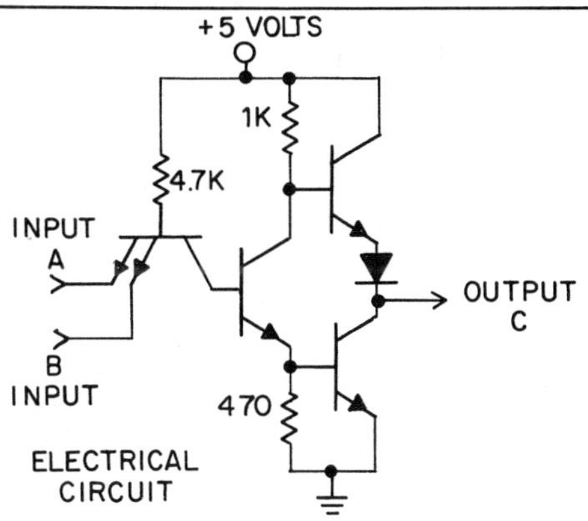

logic 1 = 2.5 - 5 volts

logic 0 = 0 - 0.4 volt

INPUT		OUTPUT
A	B	C
1	1	0
1	0	1
0	1	1
0	0	1

LOGIC TABLE

Figure 4.1 A Basic Integrated Circuit Logic Element Circuit Diagram with a Logic Table Description.

Mnemonics, used to name a logic line, will be drawn along the logic line to identify it. Mnemonics, drawn inside of a logic element adjacent to the inputs or outputs, refer only to the designation of the logic line internal to the logic element. These variables are called INTERNAL RESERVED VARIABLES. This convention allows the definition of a logic element independent of the definition of the interconnecting logic lines external to the logic element. For example, a logic element having three inputs, designated by the manufacturer as A, B and C and one output Q', is to be renamed for consistency with other terminology by replacing A with B, B with C, C with HOLD and Q' with HOUT. The manufacturer's circuit is called μ LOG 6605 and is to be renamed HLD1 to specify that the basic function is a HOLD of an operation on B and C and it is of type 1 in the system.

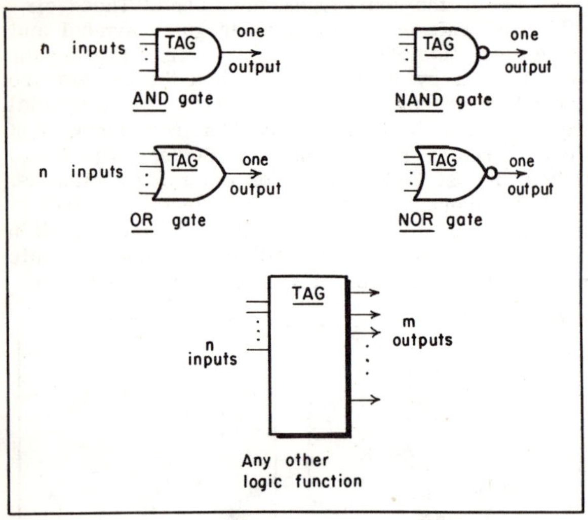

Figure 4.2 Logic Circuit Symbols.

The logic circuit symbol and the definition are given in Figure 4.3 to illustrate one use of the internal reserved variables to describe this renaming.

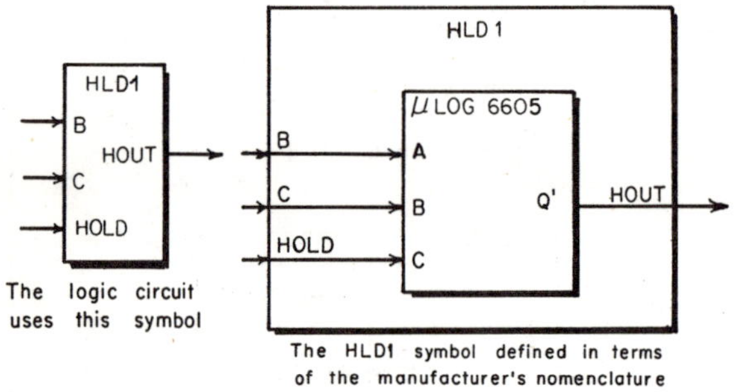

Figure 4.3 An Illustration of the Use of Internal Reserved Variables.

4.2 THE SYNTHESIS PROCESS

The synthesis of a logic circuit, in the form of logic gates, is achieved by a sequence of conversions between the machine description languages, as described in Figure 4.4. The arrows indicate the direction of the easiest conversion, and the words in parentheses characterize each description's advantage. These conversions, only briefly described in Chapter 3, are described in greater detail in this chapter. The guiding concept in the synthesis process is that the cost of a logic circuit is related to its complexity and reasonable steps to reduce that complexity should be included. A general reduction procedure, however, is insufficient because cost in integrated circuits may depend on particular restrictions involving the number of inputs to a gate or the number of gates in a single package. Therefore, the conversion and reduction procedures covered are described with flexibility in mind.

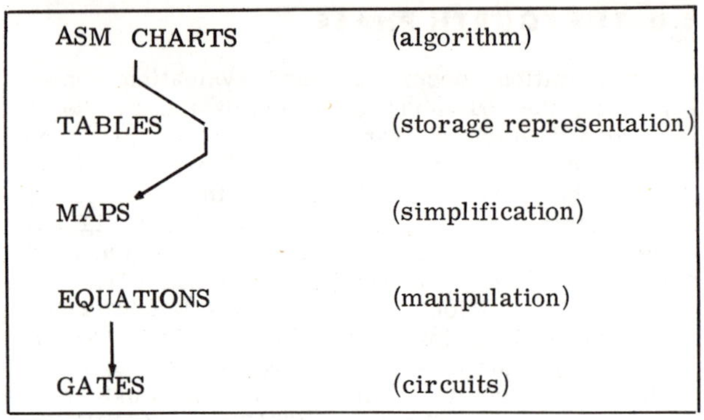

Figure 4.4 Conversion Ease and Advantages for Machine Descriptions.

4.3 EQUATION-TO-GATE CONVERSIONS

In this section the gate symbols are used to represent Boolean functions. The conversion from equation to gates is described by simple examples which demonstrate the flexibility in the process.

The gates required to synthesize an equation are determined by dividing the equation into groups of common functions where each group is all AND, NAND, NOR or OR. For example, the equations in Figure 4.5 have only one type of function.

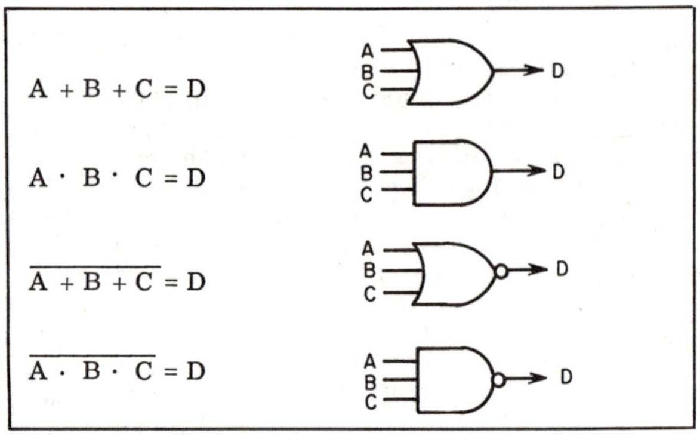

Figure 4.5 Single-Function Examples.

The equations in Figure 4.6, however, have more than one type of function and therefore must be divided into groups of common functions as indicated by the parenthesis. The resultant gate equivalents are then shown for each example. In this manner any equation can be represented by an appropriate string of gates. However, some of the four types of gates may be unavailable. The equation must then be changed to an equivalent equation which uses the gates available by employing some Boolean equivalence relations.

$D = (A \cdot B) + C$

$E = \overline{\overline{A} + (C \cdot B)} + \overline{D}$

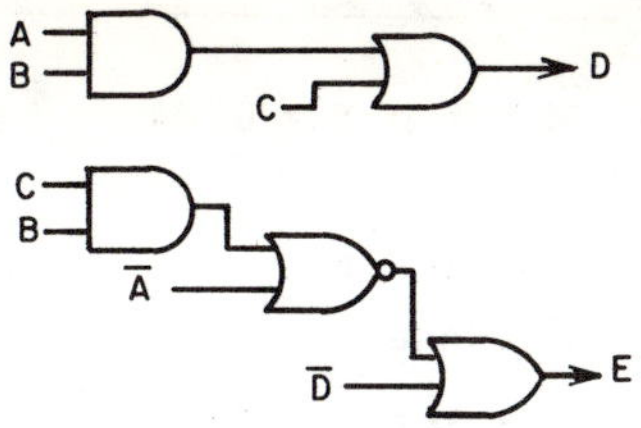

Figure 4.6 Multiple-Function Examples.

Figure 4.7 lists all the Boolean equations which are relevant to the general manipulation and understanding of Boolean equations. The postulates and single-variable theorems are almost obvious. The remaining theorems are not obvious but can be proven.[2] There are three relations which are the most useful for gate manipulations; distribution, involution and DeMorgan's theorem. Distribution is essentially equivalent to factoring. Involution and DeMorgan's theorem work together to interchange AND and OR functions. In applying these laws to Boolean equations where the number of gate levels is unimportant, the least complex gate realization is found by factoring out all the common terms and working from the outermost level down to the innermost level when converting each level to the desired logic form. The application of these three rules to gate synthesis is demonstrated by the following examples:

Example 1. Synthesize $(A+B) \cdot (C+\overline{D})$ with NAND gates only.

$(A+B) \cdot (C+\overline{D}) = \overline{\overline{(A+B) \cdot (C+\overline{D})}}$ by involution

$= \overline{\overline{\overline{\overline{A} \cdot \overline{B}} \cdot \overline{\overline{C} \cdot D}}}$ by DeMorgan's law (2 times)

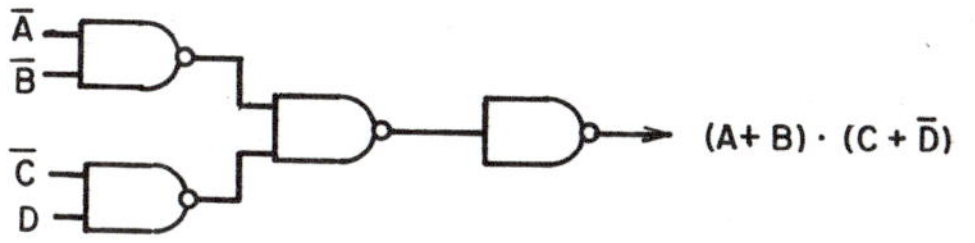

Example 2. Synthesize $A + (B \cdot (C+\overline{D}))$ with NAND gates.

$A + (B \cdot (C+\overline{D})) = \overline{\overline{A} \cdot \overline{B \cdot \overline{\overline{C} \cdot D}}}$ by DeMorgan's law (2 times)

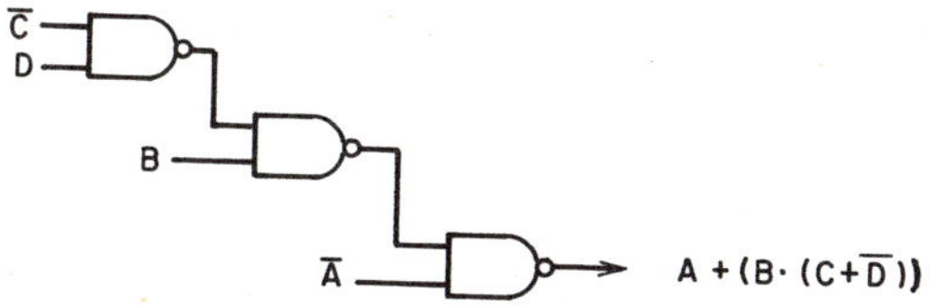

Example 3. Synthesize $A + (B \cdot (C \cdot \overline{D}))$ with 2 input NOR gates.

$A + (B \cdot (C \cdot \overline{D})) = \overline{\overline{A + (B \cdot (C \cdot \overline{D}))}} = \overline{\overline{A + \overline{\overline{B} + \overline{\overline{C} + D}}}}$

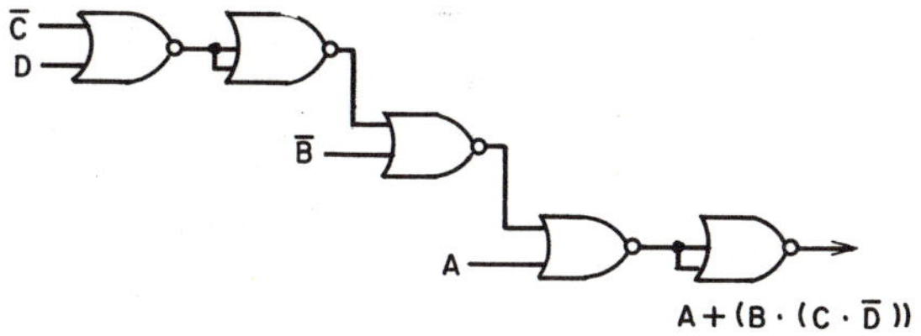

Example 4. Synthesize $A + (B \cdot (C+\overline{D}))$ with two levels of NAND logic.

$A + (B \cdot (C+\overline{D}) = A + ((B \cdot C)+(B \cdot \overline{D}))$ by distributive law

$= A + (B \cdot C)+(B \cdot \overline{D})$ by associative law

$= \overline{\overline{A} \cdot \overline{B \cdot C} \cdot \overline{B \cdot \overline{D}}}$ by DeMorgan's law

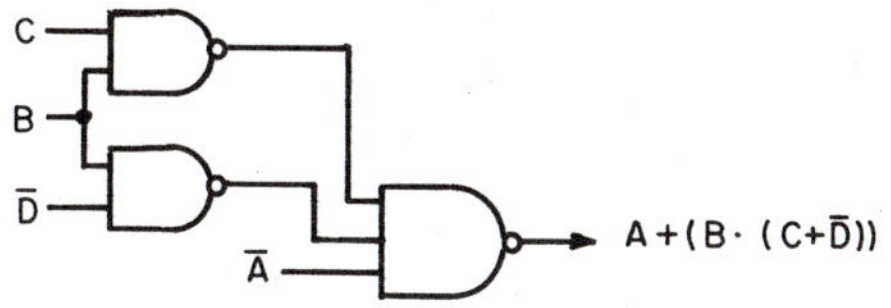

Example 5. Synthesize $(A \cdot B \cdot C \cdot \overline{D} \cdot \overline{E})+\overline{F}+\overline{G}+H+I$ with 2 input NOR gates and no more than 4 input NAND gates.

$\overline{\overline{A \cdot B \cdot C \cdot \overline{D} \cdot \overline{E}} \cdot F \cdot G \cdot \overline{H} \cdot \overline{I}}$

by DeMorgan's law

$\overline{\overline{A \cdot B \cdot C \cdot (\overline{D} \cdot \overline{E})} \cdot F \cdot G \cdot (\overline{H} \cdot \overline{I})}$

by associative law

$\overline{\overline{A \cdot B \cdot C \cdot \overline{D+E}} \cdot F \cdot G \cdot \overline{H+I}}$

by DeMorgan's law

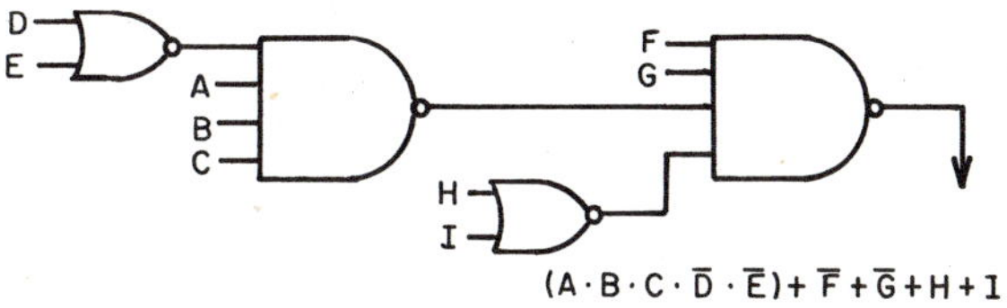

---

[2] See references at the end of this chapter for proofs of theorems.

POSTULATES

$X = 0$ if $X \neq 1$ $\qquad$ $X = 1$ if $X \neq 0$

$\overline{0} = 1$ $\qquad$ $\overline{1} = 0$

$0 \cdot 0 = 0$ $\qquad$ $1 + 1 = 1$

$1 \cdot 1 = 1$ $\qquad$ $0 + 0 = 0$

$1 \cdot 0 = 0 \cdot 1 = 0$ $\qquad$ $0 + 1 = 1 + 0 = 1$

SINGLE-VARIABLE THEOREMS

$X + 0 = X$ $\qquad$ $X \cdot 1 = X$ (Identities)

$X + 1 = 1$ $\qquad$ $X \cdot 0 = 0$ (Null element)

$X + X = X$ $\qquad$ $X \cdot X = X$ (Idempotency)

$X + \overline{X} = 1$ $\qquad$ $\overline{X} \cdot X = 0$ (Complements)

$\overline{\overline{X}} = X$ (Involution)

TWO THREE-VARIABLE THEOREMS

$X + Y = Y + X$ $\qquad$ $X \cdot Y = Y \cdot X$ (Commutative)

$X + (X \cdot Y) = X$ $\qquad$ $X \cdot (X + Y) = X$ (Absorption)

$(X + \overline{Y}) \cdot Y = X \cdot Y$ $\qquad$ $(X \cdot \overline{Y}) + Y = X + Y$

$X + Y + Z = (X + Y) + Z = X + (Y + Z)$ $\qquad$ $X \cdot Y \cdot Z = (X \cdot Y) \cdot Z = X \cdot (Y \cdot Z)$ (Associative)

$(X \cdot Y) + (X \cdot Z) = X \cdot (Y + Z)$ $\qquad$ $(X + Y) \cdot (X + Z) = X + (Y \cdot Z)$ (Distributive)

$(X + Y) \cdot (\overline{X} + Z) \cdot (Y + Z) = (X + Y) \cdot (\overline{X} + Z)$ $\qquad$ $(X \cdot Y) + (\overline{X} \cdot Z) + (Y \cdot Z) = (X \cdot Y) + (\overline{X} \cdot Z)$ (Consensus)

$(X + Y) \cdot (\overline{X} + Z) = (X \cdot Z) + (\overline{X} \cdot Y)$

n-VARIABLE THEOREMS

$\overline{(X + Y + Z + \ldots)} = \overline{X} \cdot \overline{Y} \cdot \overline{Z} \cdot \ldots$ $\qquad$ $\overline{X \cdot Y \cdot Z \cdot \ldots} = \overline{X} + \overline{Y} + \overline{Z} + \ldots$ (DeMorgan's Theorem)

$$f(X_1, X_2, \ldots X_n) = \left(X_1 \cdot f(1, X_2, \ldots, X_n)\right) + \left(\overline{X}_1 \cdot f(0, X_2, \ldots, X_n)\right)$$ (Expansion Theorem)

$$f(X_1, X_2, \ldots, X_n) = [X_1 + f(0, X_2, \ldots, X_n)] \cdot [\overline{X}_1 + f(1, X_2, \ldots, X_n)]$$

Figure 4.7 Summary of Boolean Equivalence Relations.

The desirability of one gate solution over another involves the cost of the gates; there is really no general way to determine the cost without trying some solutions. As these examples have shown, a problem may be subject to a variety of constraints which greatly affect the cost of the solution.

4.4 USING MAPS TO SIMPLIFY BOOLEAN FUNCTIONS

This section describes the use of the map description for recognizing simplified forms of Boolean equations by means of READING map SUBCUBES. Complex maps can be reduced by MAP-ENTERED VARIABLES. Reading a map is summarized by a general algorithm.

A very important theorem, thus far not covered, is absorption. Absorption summarizes the process called SIMPLIFICATION, which means the removal of redundant terms from an equation. Although the absorption rule is effective in practice, it is often difficult to identify the best way to absorb redundant variables to reach the simplest expression. This process becomes very difficult when some variable combinations are unimportant. For the job of reduction, the map turns out to be a more useful description than the equation. For example, the map makes it easy to see that the equation $(\overline{D}\cdot\overline{C}) + (A\cdot D\cdot\overline{B}) + (D\cdot\overline{C}\cdot\overline{A})$ with $(\overline{D}\cdot C) + (D\cdot B\cdot A)$ as don't care combinations is really equivalent to $A + \overline{C}$. This example will be demonstrated later.

Map Subcubes and Reading Maps

Map descriptions lend themselves to identifying minimum-literal expressions[3] for Boolean functions. The technique involves recognizing certain patterns, called SUBCUBES, in the map. In Chapter 3, a map was described as a two-dimensional representation of an n-cube. A subcube is any smaller cube contained in the n-cube. Any single point in an n-cube is a 0-cube and is represented by n variables. Each successively higher order subcube is represented by one less variable until the complete n-cube is reached with no variables required. For example, in Figure 4.8, a 3-cube with several subcubes, indicated by shading, has three variables for a 0-cube, two variables for a 1-cube, one variable for a 2-cube and zero variables for the 3-cube. Just as with any n-cube, the number of vertices in a k-subcube will be 2^k; and consequently, the number of squares in the map will be 2^k. Any combination of subcubes which when combined form a higher order subcube may be expressed with fewer variables. Thus, a simplified expression is formed by the smallest number of highest order subcubes that include the desired points. This technique is the basis for map reduction.

[3] Minimum-literal expressions have the fewest possible number of occurrences of each variable.

Figure 4.9 shows several map representations for some typical subcubes. For obvious reasons, the subcubes are often called MAP ENCIRCLEMENTS.

To see how the maps are used to simplify equations, consider the equation $Z = (A\cdot\overline{B}) + (\overline{B}\cdot C\cdot A) + C$, which may be described by one map as in Figure 4.10. Blank entries are assumed to mean 0. It is easy to see from the complete map in Figure 4.10 that two subcubes are sufficient to cover all the 1's; that is, Z will be a 1 for any point in the two subcubes. Of course, the same result could be obtained by the use of an absorption equivalence relation; maps are just a little easier. Translating the map into Boolean equations is called READING THE MAP.

The same map as shown in Figure 4.10 may be read in another way by encircling the maximum subcubes to cover all the 0's. The function Z is then the complement of this function and equivalent to the previous solution as shown in Figure 4.11. Reading the 0's of a map will often produce a simplified expression of an equation in less time than reading 1's if there are fewer 0's than 1's in the map. As shown, there is really no more information obtained by reading 0's (although it is sometimes useful when working with OR gates) and therefore reading 0's is seldomly used.

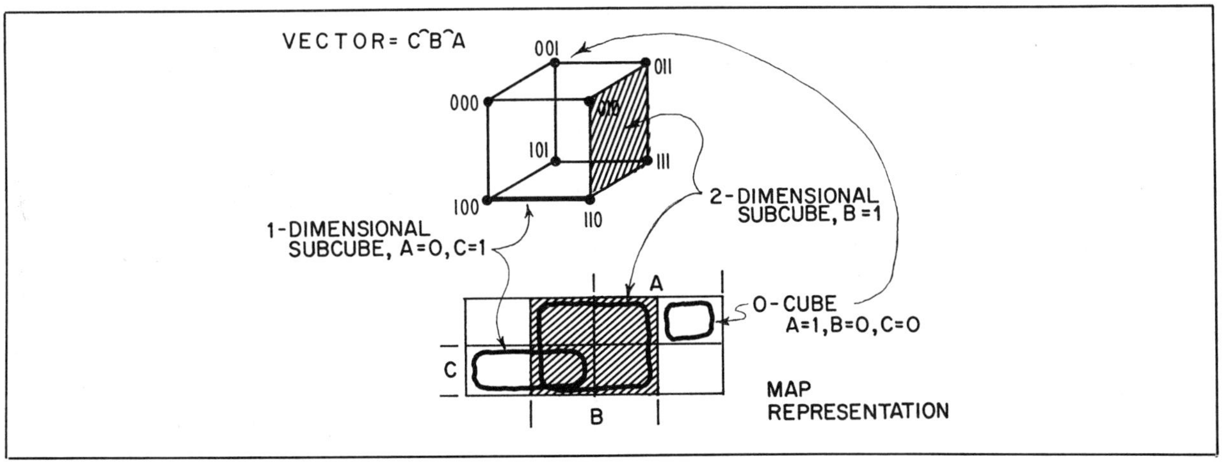

Figure 4.8 Map Subcube Representations.

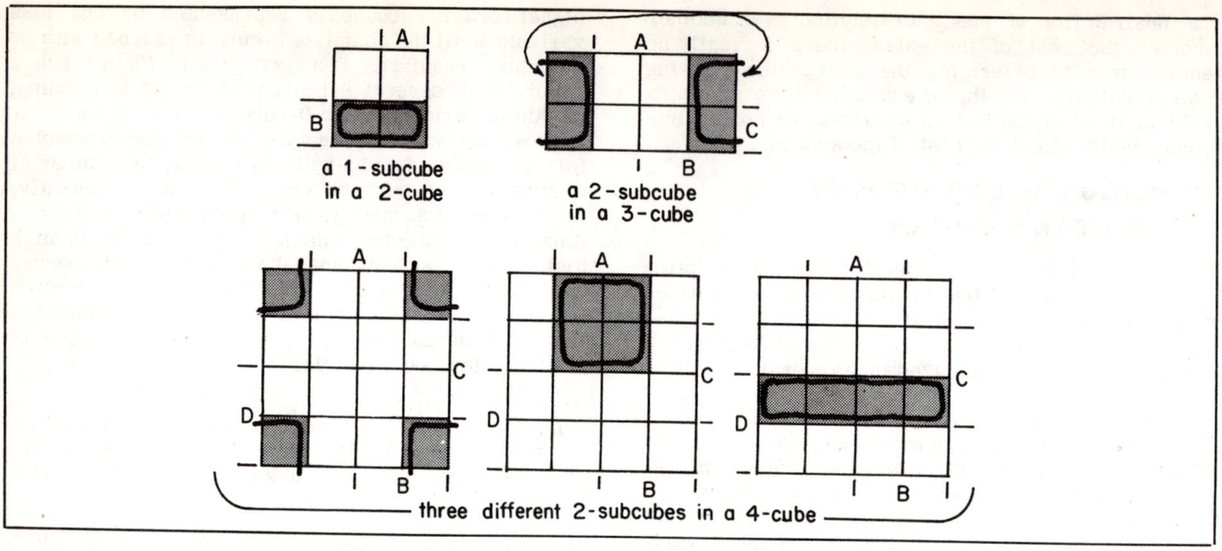

Figure 4.9 Several Subcube Representations.

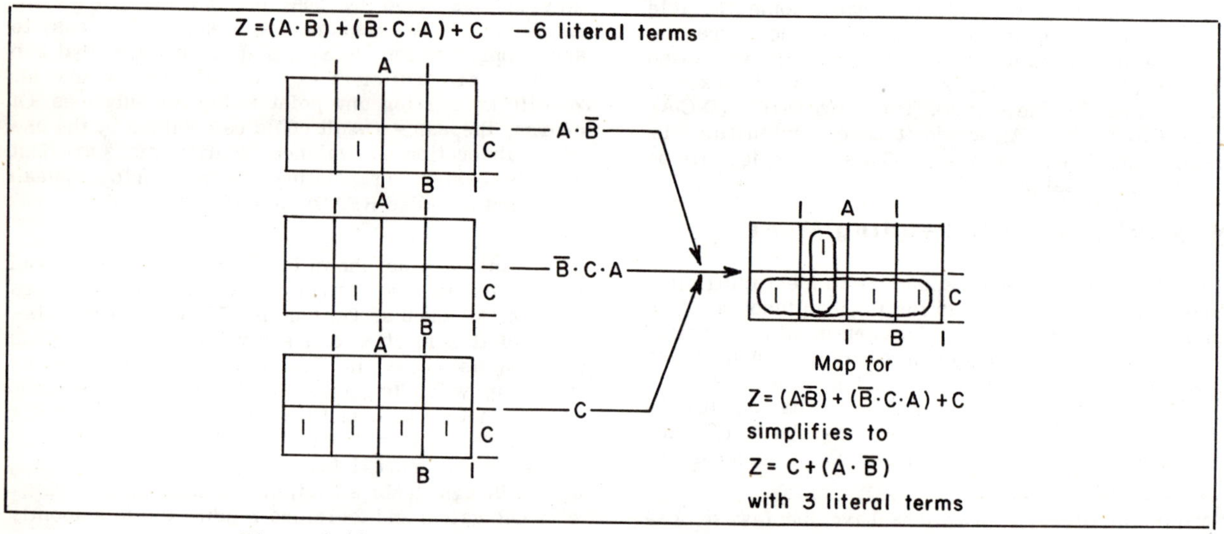

Figure 4.10 The Use of a Map to Identify the Simplified Equation.

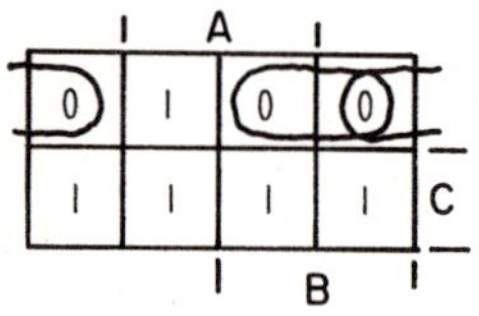

$\overline{Z} = (B \cdot \overline{C}) + (\overline{A} \cdot \overline{C})$

$Z = \overline{(B \cdot \overline{C}) + (\overline{A} \cdot \overline{C})}$

Prove that $\overline{(B \cdot \overline{C}) + (\overline{A} \cdot \overline{C})} = C + (A \cdot \overline{B}) = Z$

$Z = \overline{(B + \overline{A}) \cdot \overline{C}}$	by distributive law
$= \overline{B + \overline{A}} + C$	by DeMorgan's law
$= (\overline{B} \cdot A) + C$	by DeMorgan's law
$= C + (A \cdot \overline{B})$	by communicative law
	<u>QED</u>

Figure 4.11 Reading the 0's of a Map.

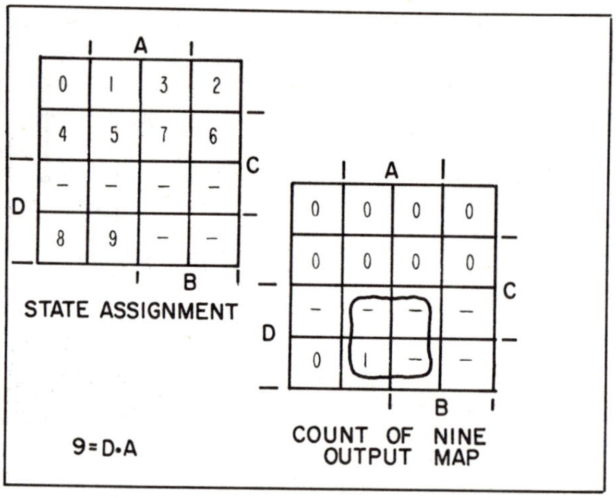

Figure 4.12 Using Don't Care States to Reduce the Literals in a 9 Detector.

Often in combinatorial logic certain input combinations either never occur or are not important in the final solution. These inputs are called DON'T CAREs and may be chosen as a 1 or a 0, whichever helps to produce the largest covering subcubes. For example, a decade counter, as described in the last chapter, uses only 10 of the 16 possible combinations of the four variables. The remaining six states are don't care states for decoding of various count outputs. In Figure 4.12, dashes show the position of the don't cares in a map of the counter states. The largest cover using the don't cares to decode the count of 9 has the equation D·A.

In the introduction to this section, another problem was mentioned using don't care entries which had a very simple solution, $A + \overline{C}$. Using maps and don't cares, this solution is obvious. The problem is

$$(\overline{D}\cdot\overline{C}) + (A\cdot D\cdot\overline{B}) + (D\cdot\overline{C}\cdot\overline{A})$$

$$\text{with don't care} = (\overline{D}\cdot C) + (D\cdot B\cdot A)$$

The map for this function is shown in Figure 4.13 from which the solution $A + \overline{C}$ is easily read.

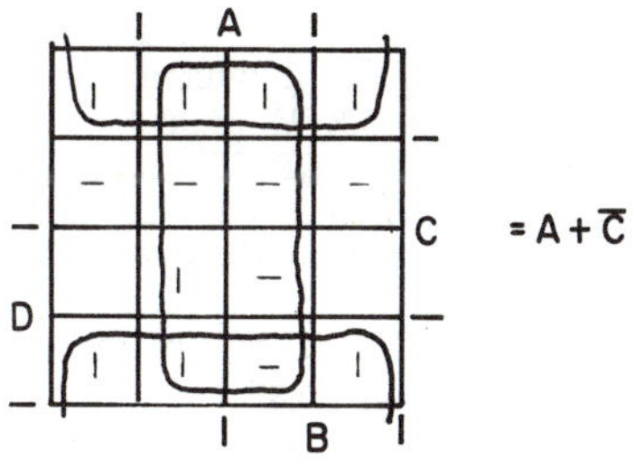

Figure 4.13 Example Reduction.

Reducing the Required Map Size with Map-Entered Variables

Maps provide a convenient display of an equation for identifying the simplified terms. However, for more than six variables in one problem, the map itself becomes very unwieldy to handle manually. This section describes an addition to the map description which reduces the map size required in complex problems. This addition makes maps useful in most practical problems.

The number of map dimensions required to represent a problem can be less than the number of problem variables. Normally, the number of map dimensions in a complete map equals the number of problem variables. A smaller map can contain the same information as the complete map when the additional information is provided by map-entered variables. MAP-ENTERED VARIABLES are other symbols on the map in addition to the 1, 0 and don't care so far mentioned. The map in Figure 4.14 illustrates the meaning of map-entered variables. The complete map on the left describes a function X = (B·C) + (C·A). The reduced map on the right also describes X = (B·C) + (C·A), but it uses A as the map-entered variable.

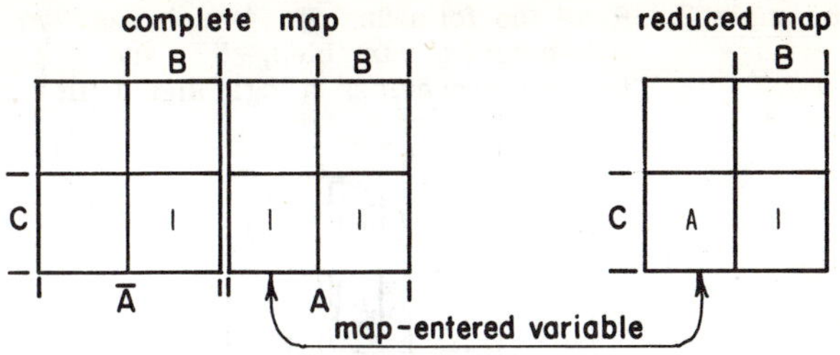

Figure 4.14 A Reduced Map Example.

The A in the reduced map means that a 1 occurs in the same location in the A field of the complete map and a 0 occurs in the corresponding location in the $\overline{A}$ field. The 1 in the reduced map means that a 1 occurs in both fields of A in the complete map, which makes the variable A a don't care in this position of the reduced map. In any example, the number of map-entered variables plus the number of map variables will equal the number of problem variables. Thus, the map dimension can be reduced by the number of map-entered variables.

It is more important to be able to read a reduced map than it is to be able to reduce a map. The last example demonstrated a simple reduction from three to two dimensions. Continued reductions can reduce the map to a 0-cube which contains a single equation. The reduction process becomes more complicated when symbols appear in the maps to be reduced. Furthermore, it can be shown that it is difficult to obtain the simplest equation by this method. For these reasons, the reduction process is best suited for compacting a description rather than obtaining an equation. The remainder of this chapter describes ways to form maps with map-entered variables, using the ASM chart. The skill to be learned is how to read such a map using map encirclements.

A map containing map-entered variables is read in two steps. Step one is to read all the 1's counting all map-entered variables as 0. Step two is to take each map-entered variable separately and find an encirclement using previously circled 1's as don't cares. These steps can be completed on a single map by mentally changing the variables. In example 1, the first step is to let the map-entered variable A be 0 when encircling 1's, which yields $\overline{C}$. The second step is to let the 1's be don't cares when finding the largest encirclement including A, which is B. This encirclement is ANDed with the map-entered variable A. Thus, the encirclement of A would be read A·B· The OR of all encirclements gives the final solution $Z=\overline{C}+(A\cdot\overline{B})$. Additional examples follow.

Example 1. Read the following map for Z with the map-entered variable A.

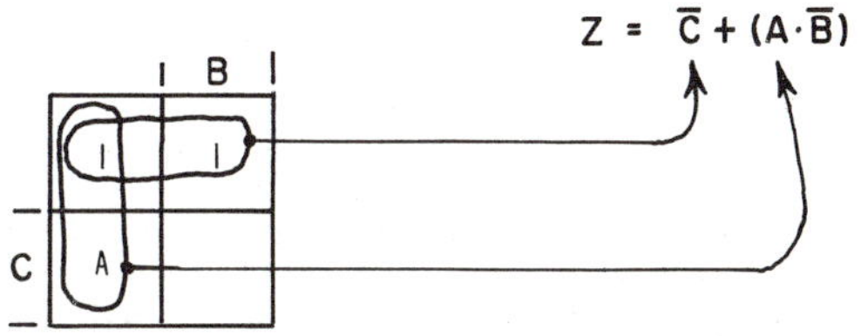

Example 2. Read the following map which has two map-entered variables, YZ and YX. The map describes a logic signal HRT.

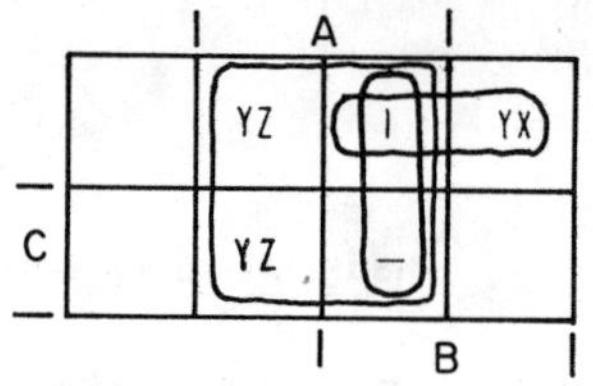

$$HRT = (A \cdot B) + (YZ \cdot A) + (YX \cdot \overline{C} \cdot B)$$

Example 3. The following map has three map-entered variables, X, Y and Z. The encirclements are left out as an exercise for the reader.

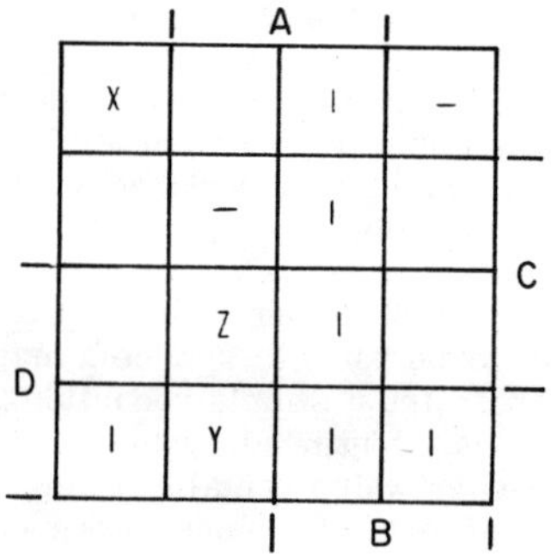

$$MAP = \underbrace{(A \cdot B) + (\overline{A} \cdot \overline{C} \cdot D)}_{\text{1's alone}} +$$

$$\underbrace{(X \cdot \overline{A} \cdot \overline{C}) + (Z \cdot C \cdot A) + (Y \cdot \overline{C} \cdot D)}_{\text{map-entered terms}}$$

Example 4. The following map can be read two ways:

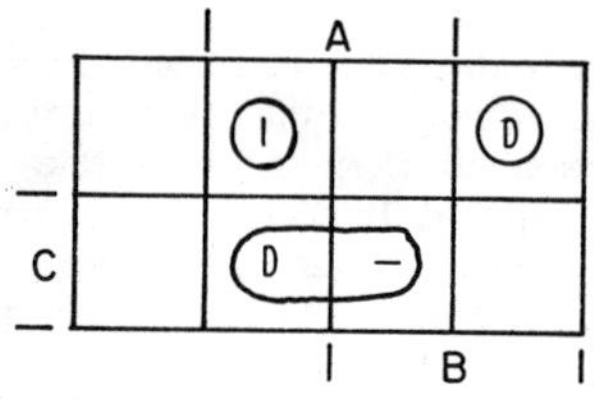

$$(A \cdot \overline{B} \cdot \overline{C}) + (D \cdot C \cdot A) + (D \cdot \overline{A} \cdot \overline{C} \cdot B)$$

OR

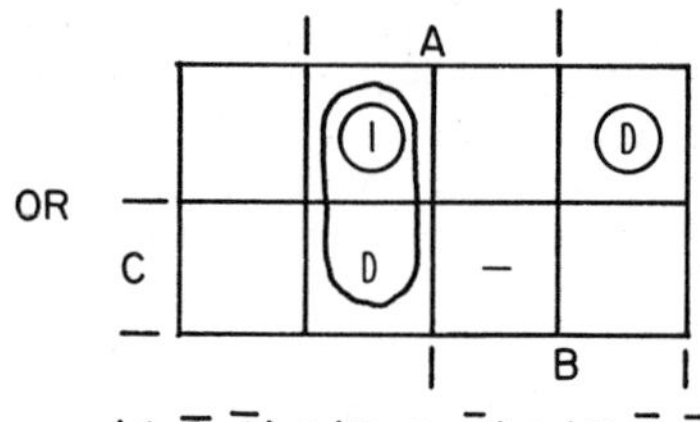

$$(A \cdot \overline{B} \cdot \overline{C}) + (D \cdot A \cdot \overline{B}) + (D \cdot \overline{A} \cdot \overline{C} \cdot B)$$

A map-entered variable may be a single term or a Boolean expression. Whenever two adjacent squares contain expressions with a common variable, or a variable and its complement, some additional simplification might be realized if the variable in question is made a map variable. Of course, this choice doubles the map size and, unless minimization is essential, such an expansion may not be warranted. Reading a map with map-entered variables always yields a valid solution and, observing the above caution, it also yields as minimum a solution as any size map. The value of map-entered variables is seen when complex sequential machines can be minimized using simple maps. More complex minimization techniques are rarely needed. In the rest of this book, various examples will show that map-entered variables are a natural consequence of logic design using ASM charts.

A Map-Reading Algorithm

All the map-reduction techniques for single outputs may may be summarized by Algorithm M for obtaining the minimum number of maximum size subcubes.

ALGORITHM M. Find the minimum-literal expression from a map which may include seldomly used variables and don't care entries. The map dimension is n, and k is a counting integer.

M1. Consider all seldomly used variables as 0's and set k =0.

M2. Encircle all the uncircled 1's which can be included in only one k-subcube and cannot be included in any (k+1)-subcubes. These are the ESSENTIAL TERMS. Encircle the remaining 1's which cannot be included in a (k+1)-subcube. Use don't cares where helpful.

M3. Set k = k + 1. If all 1's are circled, then go to M4; otherwise, go to M2.

M4. Set one uncircled seldomly used variable to 1 and consider all other seldomly used variables as 0 and all other circled 1's as don't cares. If all seldomly used variables are circled, go to M5; otherwise, set k = 0 and go to M2.

M5. Read each subcube, including any circled seldomly used variable in the expression.

M6. For maps with no seldomly used variables, read the 0's and compare with M5.

Other Methods of Finding Gate Solutions

There has been considerable study in switching theory concerning better methods to find the best gate solution for a given equation set. All of these techniques, however, are extensions of the methods described in this section and consist of two major operations: (1) identifying all the possible subcubes, often called MINTERMS, and (2) selecting the minimum set of subcubes that cover the desired function. The Quine-McClusky table technique is one well-known method but is quite lengthy. The algorithms by Roth, as reported in Miller, are also very powerful but are more mathematical. Roth's sharp (#) operation finds all the subcubes in one algorithm. Roth's algorithms are TOPOLOGICAL (based on the state space), while McClusky's algorithms are TABULAR (based on Boolean equations). References to works by these authors and others are given at the end of the chapter. Further discussion is not required for the purposes of this book because these techniques are generally computer oriented rather than concept oriented.

4.5 OUTPUT FUNCTION SYNTHESIS

In this section, it is shown by a simple example that the ASM chart outputs result in combinatorial output maps with map-entered variables which can be synthesized as in the preceding section. Figure 4.15 describes a four-state machine, with several state and conditional outputs. The state-assignment map gives the location of the states on the map of the state variables, A and B. The qualifiers are the map-entered variables. IHZQ is a state output which occurs in state ⓐ and state ⓒ. The map for IHZQ, then, has a 1 entered in these states to represent the output during these states. IHQT occurs in state ⓑ only. HR and HMF are both conditional outputs from one state and state outputs from another. The condition required for HMF in state ⓑ is YZ = 1; therefore, a YZ is entered into state ⓑ of the HMF map. The condition for HR in state ⓐ is YT = 0; therefore, a $\overline{YT}$ is entered into state ⓐ of the HR map. The output equations for HMF and HR are read from the maps using the techniques of map-entered variables, yielding $HMF = A \cdot (B + YZ)$ and $HR = \overline{A} \cdot (\overline{YT} + B)$. States which are not used on the map are don't care states in every output map. If an output is of the L-type, as defined in Section 2.2, the output equation is complemented.

4.6 MULTI-OUTPUT SIMPLIFICATION

In this section, some additional gate simplifications are described for multiple outputs from a common set of variables, such as the state variables. The basic concept is to isolate common terms shared by several outputs and use these terms to save gate inputs.

Common terms are often easy to identify in the Boolean equations. For example, the Boolean equations

$$H1 = \boxed{(A \cdot C)} + (B \cdot \overline{D}) + (C \cdot B)$$
$$H2 = \boxed{(A \cdot C)} + (C \cdot \overline{B}) + (B \cdot D)$$

Common Terms

obviously have $A \cdot C$ as a common term. The gate implementation for H1 and H2 can share this common term, as shown in Figure 4.16.

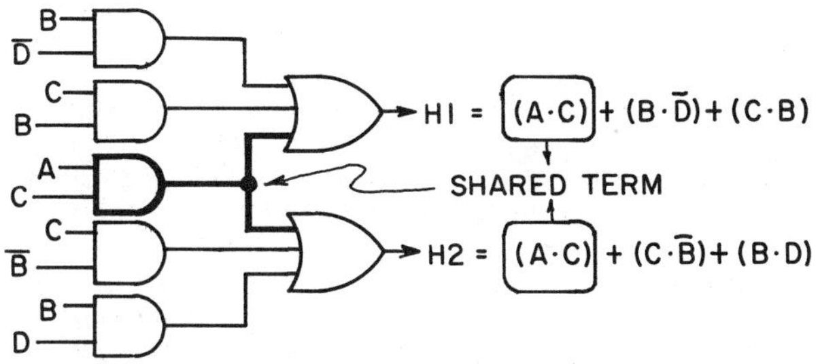

Figure 4.16 A Common OR Term Saving Gate Inputs.

Another example of common terms in two equations is found in the following output equations:

$$H1 = (A \cdot \boxed{B \cdot C \cdot \overline{D})} + (\overline{A} \cdot \overline{B})$$
$$H2 = (\overline{A} \cdot \boxed{B \cdot C \cdot \overline{D})} + (\overline{C} \cdot D)$$

Common Terms

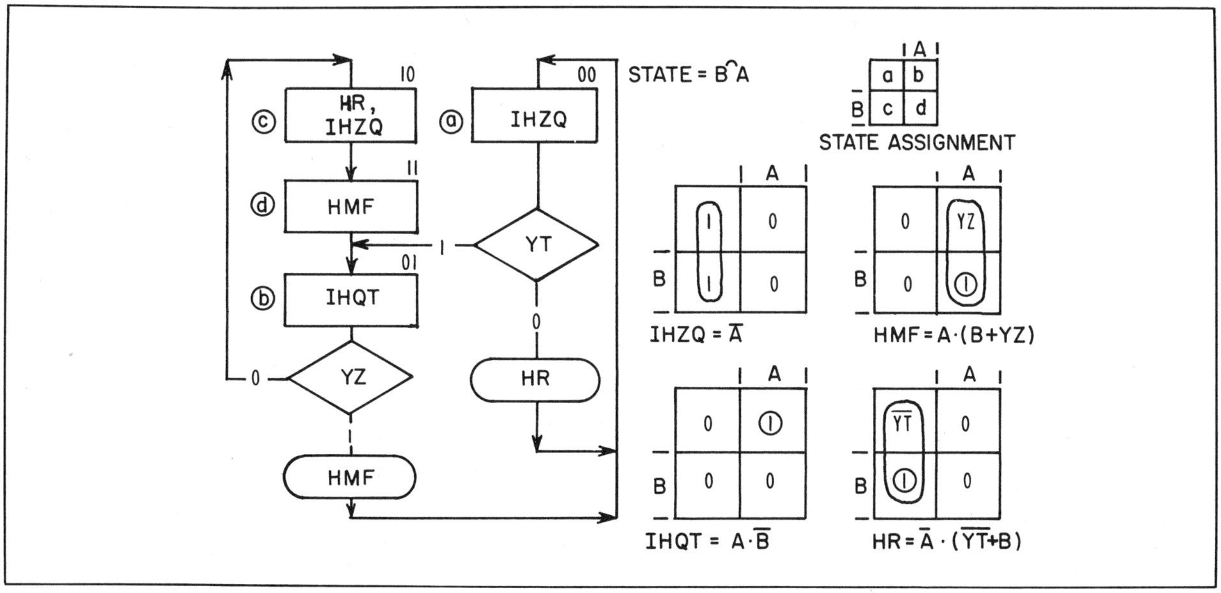

Figure 4.15 The Synthesis of Several Outputs of a Conditional Output State Machine.

These common terms are in an AND function; and when isolated in a common gate, they can also save gate inputs, as illustrated in Figure 4.17.

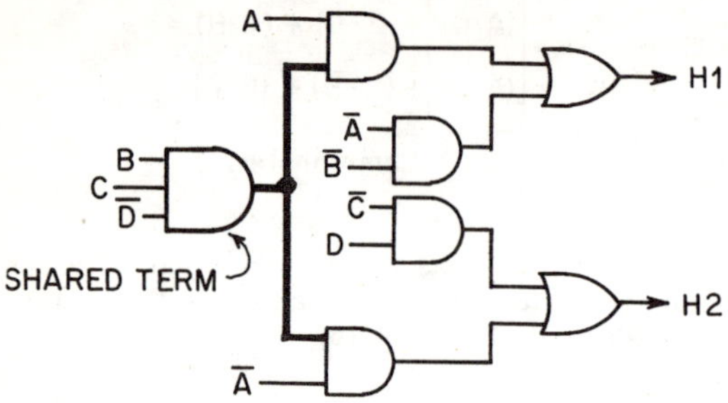

Figure 4.17 A Common AND Term Saving Gate Inputs.

Common OR or common AND terms can be recognized from the map representations as well as from the equations. The OR terms appear as common 1 outputs in two or more maps, as in Figure 4.18. Sometimes redundant covers are made just to form common OR terms. The common AND terms are a little less obvious. They appear on the map as subcubes, covering groups of outputs in both maps but requiring additional restrictions by indicated map variables as illustrated in Figure 4.19. Common AND and OR terms can be utilized to save a great many gates in complex multi-output functions. The exact solution used may depend on many factors which are hard to express in a simple rule. This section has just introduced the techniques for recognizing common terms.

4.7 NEXT-STATE FUNCTION SYNTHESIS

In this section the process for finding the gate implementation for the next-state equation is discussed for any ASM chart or tabular description of a state sequence with a given set of state codes. The process first covers state machines using no state storage elements; then it covers forming the input function for various kinds of flip-flops used as the state storage elements. The basic concepts introduced include the RS, T, D and JK FLIP-FLOPS, the CLOCK INPUT, the PARTIAL NEXT-STATE EQUATION and the TRANSITION TABLE.

Synthesizing Flip-Flops (Unclocked State Machines)

Flip-Flops are excellent examples for discussing unclocked state machines.[4] The RS flip-flop is the simplest flip-flop and may be described by the ASM chart in Figure 4.20, which shows that there are two essential states, ⓐ and ⓑ. In this example, as in all unclocked state machines, the state time is determined by the qualifier inputs. If one or the other of the two inputs is equal to 1, the state machine is forced into one of the two states. The set input causes the IHFX output to equal 1. If, however, both

[4] Even though flip-flops often have clock inputs, they are still unclocked in the sense that the state time is determined by inputs alone. A clocked system uses these flip-flops with clock inputs. See Section 3.5 and the discussion of JK flip-flops in this section.

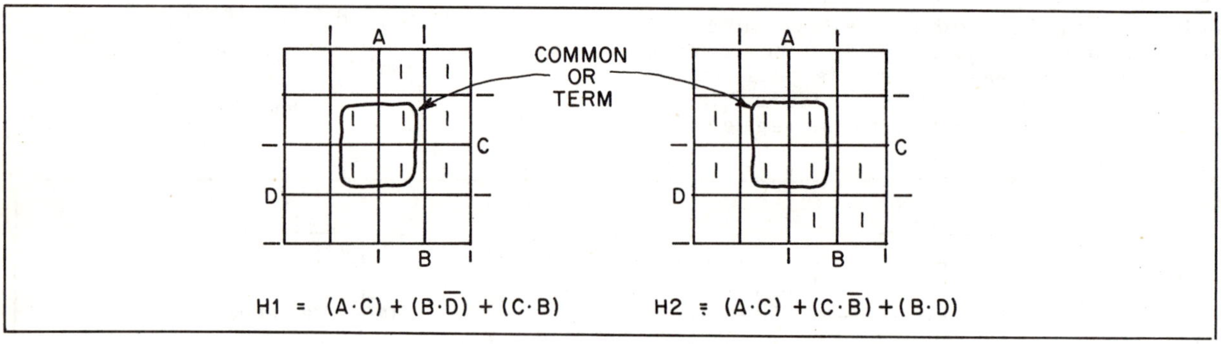

Figure 4.18 Recognizing a Common OR Term.

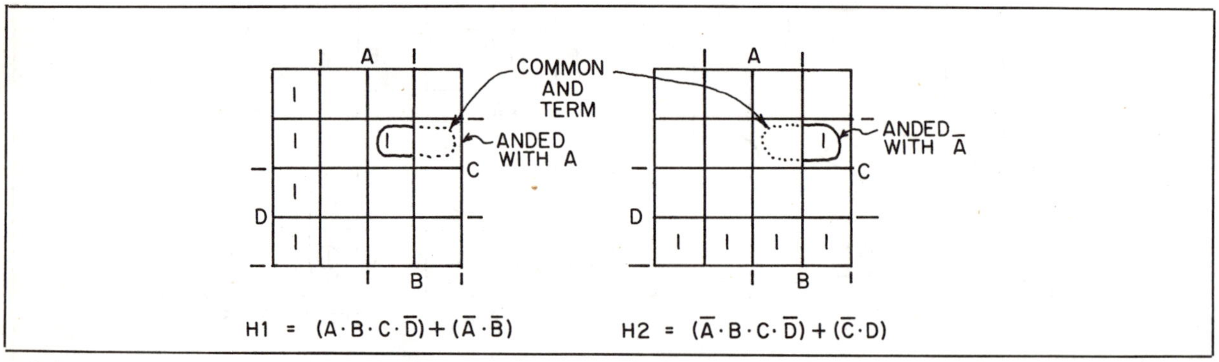

Figure 4.19 Recognizing Common AND Terms.

YS and YR equal 1, the machine races between states making the present state uncertain until one of the two inputs goes to 0. For this reason, the state of an RS flip-flop is generally undefined when both R and S are equal to 1.

The RS flip-flop is the easiest of the flip-flops to synthesize. The simplified next-state equation can be found by constructing a next-state map for the ASM chart description in Figure 4.20. The map can be made from the table description of the ASM chart or directly from the ASM chart by reading it as a kind of table. Figure 4.21 gives a tabular description, the map and the next-state equation for the ASM chart given in Figure 4.20.

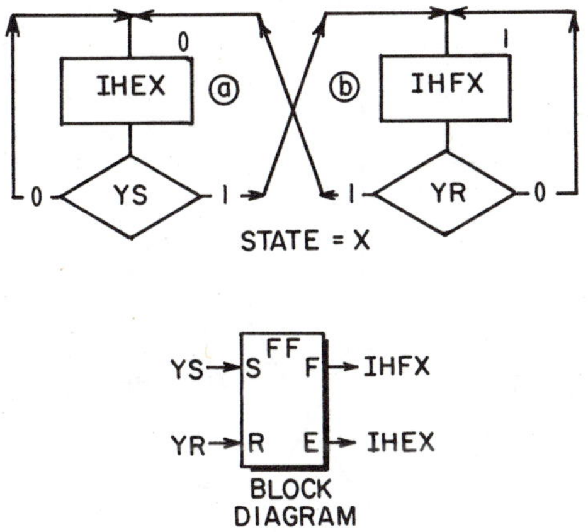

IHEX = [E OUTPUT = 1]

IHFX = [F OUTPUT = 1]

YS = set input

YR = reset input

Figure 4.20 The ASM Chart for an RS Flip-Flop.

PRESENT STATE					NEXT STATE
INPUT		STATE	OUTPUT		
YR	YS	X	IHEX	IHFX	X
–	0	0	1	0	0
–	1	0	1	0	1
0	–	1	0	1	1
1	–	1	0	1	0

TABLE

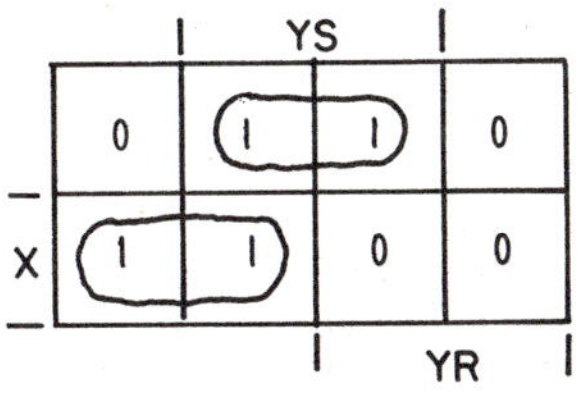

$X \leftarrow (YS \cdot \overline{X}) + (\overline{YR} \cdot X)$

NEXT-STATE MAP

Figure 4.21 Table, Map and Equation Description of the Next State of an RS Flip-Flop.

The key step in synthesizing the next-state equation is making the delay function an immediate function. Initially, the delay function was used to indicate the next-state behavior of the function described. In this example and other unclocked machines the state time becomes negligible compared to any input time, and thus the delay function becomes an immediate function. The next-state equation in Figure 4.21 can be made an immediate function by replacing the delay operator, ← , with an equal sign, =. The resulting Boolean equation can be converted to gates as shown in Figure 4.22.

$X = (YS \cdot \overline{X}) + (\overline{YR} \cdot X)$ $\qquad IHFX = X$

$X = \overline{\overline{YS \cdot \overline{X}} \cdot \overline{\overline{YR} \cdot X}}$ $\qquad IHEX = \overline{X}$

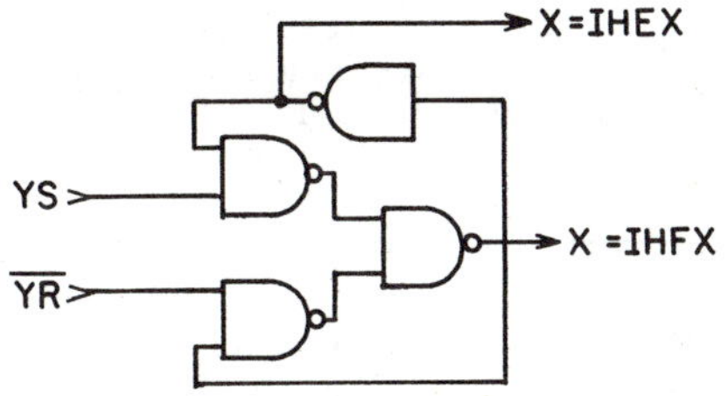

Figure 4.22 A Direct Implementation of the Next-State Equations for the RS Flip-Flop.

The RS flip-flop is further simplified by assuming the simultaneous input of YS and YR = 1 will never occur. This choice is prompted because the rapid changing of state indicated by the ASM chart in Figure 4.20 for these same conditions is a useless behavior for a circuit in a logic system. Figure 4.23 gives the next-state map and equation for the RS flip-flop when the input condition YS and YR = 1 is called a don't care. Several possible implementations are given in Figure 4.24.

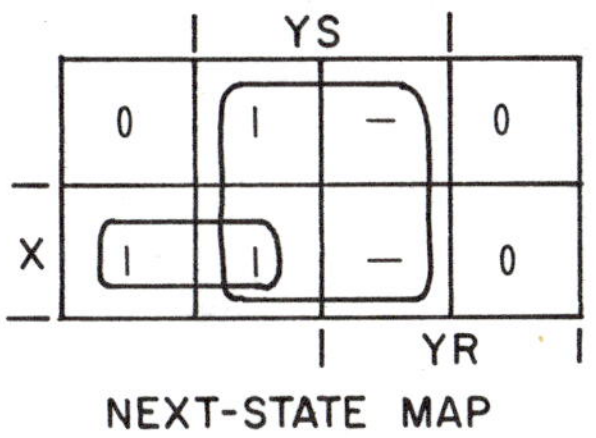

NEXT-STATE MAP

$X \leftarrow YS + (X \cdot \overline{YR})$, $\overline{X} \leftarrow YR + (\overline{X} \cdot \overline{YS})$

Figure 4.23 The Next-State Map and Equation for the RS Flip-Flop Using YR and YS = 1 as a Don't Care.

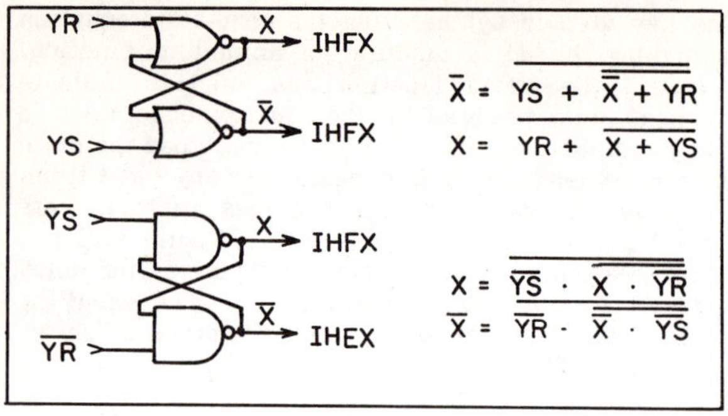

Figure 4.24 The Implementations of an RS Flip-Flop Using YR and YS = 1 as a Don't Care Input.

Each of these examples is an RS flip-flop. The symbol for the RS flip-flop, to represent one of the circuits in the previous discussion, is a box with the tag FF, inputs $\overline{R}$ and $\overline{S}$ and outputs F and E,[5] as shown in Figure 4.25. For the purposes of this book, a more specific tag is not required.

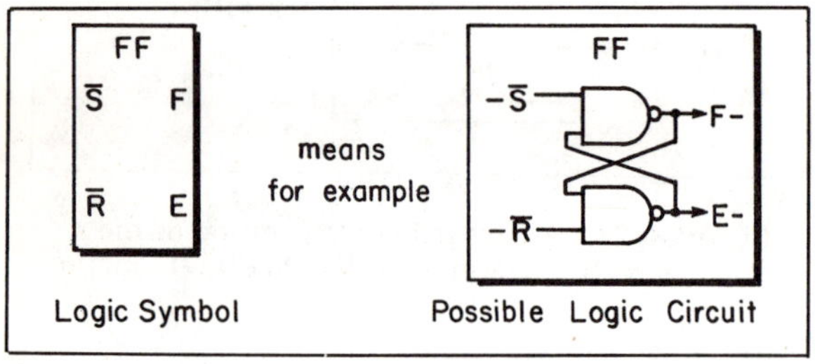

Figure 4.25 An RS Flip-Flop Symbol.

The D, T and JK flip-flops are three similar state machines which often are elements of more complex machines. Unlike the RS flip-flop, these flip-flops have an input called clock, which is used to establish the state time, and a next state which is specified for all input combinations. Figures 4.27, 4.28 and 4.29 show the ASM charts for these three flip-flops. The similarity is evident. In all three the outputs are the same in states ⓐ and ⓑ, and in states ⓒ and ⓓ. When any flip-flop is in state ⓑ or ⓓ, the clock input, YC will cause a change in the output when it goes to a 0. Each flip-flop has a single qualifier setting up the transition from ⓐ to ⓑ, or from ⓒ to ⓓ. It is the small differences in these qualifiers that separate the behavior of the flip-flops.

The delay flip-flop, or D flip-flop, has a clock input and a D input. At each 1 to 0 transition of the clock, the D flip-flop takes the input during the 1 level of the clock and holds this level on the output until the next 1 to 0

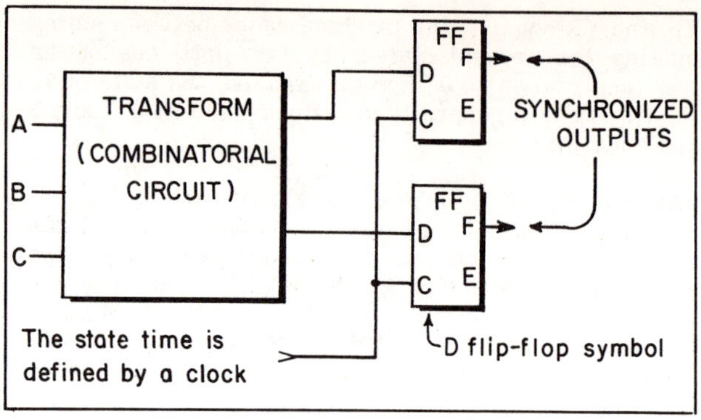

Figure 4.26 D Flip-Flop as a Delay Synchronizer.

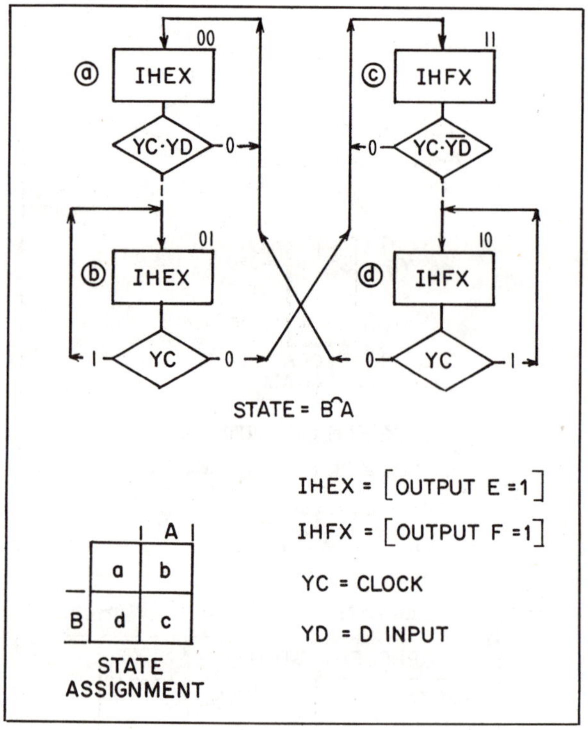

Figure 4.27 An ASM Chart for a Possible Clocked Delay Flip-Flop.

clock transition. In this manner, it delays a change at the input until the end of a state time, which is defined by the clock. Figure 4.26 shows the use of D flip-flops to synchronize the outputs of a combinatorial circuit with the state times of a clock. The ASM chart in Figure 4.27 shows that the D input, called YD, and its complement, called $\overline{YD}$, are used to set up the D flip-flop for the proper transition while the clock is set to 1. Once states ⓑ and ⓓ are reached in the setup, the next state is determined by the clock alone. This behavior means the D input must be correct while the clock is set to 1. Even very short pulses on the D input during this time might change the flip-flops's state. Flip-flops with this behavior are often called "one's catchers" and are quite useful in a system when used with proper care.

[5] The use of F and E flip-flop outputs rather than the sometimes more conventional Q and $\overline{Q}$ was chosen because the two flip-flop outputs are two separate outputs rather than one signal and its complement as implied by Q and $\overline{Q}$. A case in point is the RS flip-flop in Figure 4.24; E ≠ $\overline{F}$ when $\overline{S}$ and $\overline{R}$ = 0.

The toggle flip-flop, or the T flip-flop for short, has an ASM chart as shown in Figure 4.28. Both states (a) and (c) have the same qualifier, YC·YT. While YT=1, the output will change state, or toggle, on each clock. While YT=0, the output will remain unchanged. This flip-flop is a one's catcher.

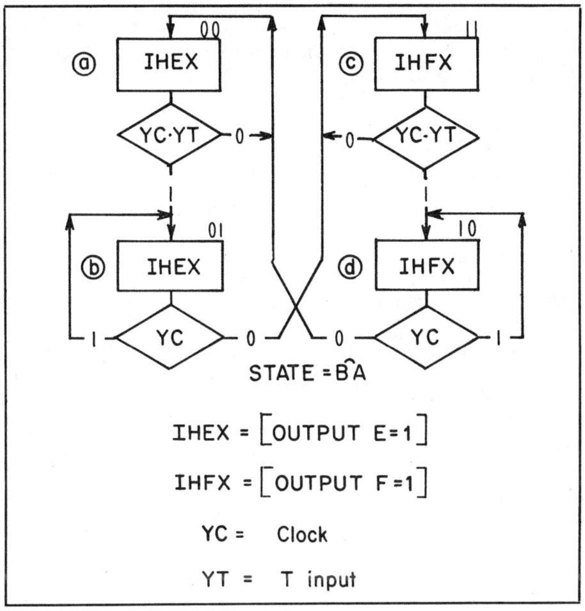

Figure 4.28 The ASM Chart for a T Flip-Flop.

The JK flip-flop, shown in Figure 4.29, has separate inputs, J and K, to set up the next state from states (a) and (c). This feature makes the JK more general than either the D or T. The JK flip-flop can be viewed as a clocked RS with its behavior defined for all input combinations. When both J and K are equal to 1, the flip-flop toggles on each clock; otherwise, the J input acts like a clocked SET, and the K input acts like a clocked RESET. This JK is also a one's catcher.

The syntheses of a D, T or JK flip-flop as just described can proceed directly from the ASM chart. The two functions to synthesize are the next-state function and the output function. A complete design will be done for the JK fkip-flop using the state assignment shown in Figure 4.29. (State assignment is covered in Section 4.8.)

First, the output function is found by making a map using the state variables as map variables and entering a 1 for each state giving the desired output. A separate map is made for each output. Looking at Figure 4.29, there are two outputs, IHEX and IHFX. The output IHEX occures in states (a) and (b), therefore, a 1 is entered in these state positions as shown in Figure 4.30. Reading this map yields IHEX=$\overline{B}$. Repeating this process for output IHFX yields IHFX=B, as shown in Figure 4.30.

Second, the next-state function is synthesized by deriving the partial next-state function for each state variable. These two functions can be found in the same manner as the next-state function for the RS flip-flop, or the functions can be found directly from the ASM chart by using maps and map-entered variables. The direct method will be taken. Starting with a map of the states using the state variables as map variables, an entry is made for each state by looking at the behavior of a single state variable for every link path out of the state. The conditions for all link paths that lead to a next state of 1 are entered as map-entered variables. When every state is completed for a single state variable, the resultant map is read to yield the partial next-state equation for that state variable.

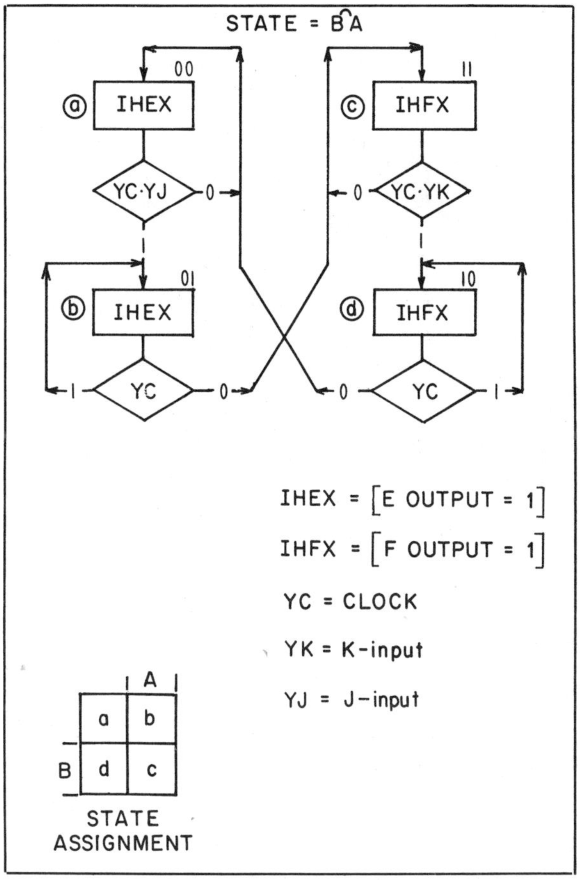

Figure 4.29 An ASM Chart Description of a Clocked JK Flip-Flop.

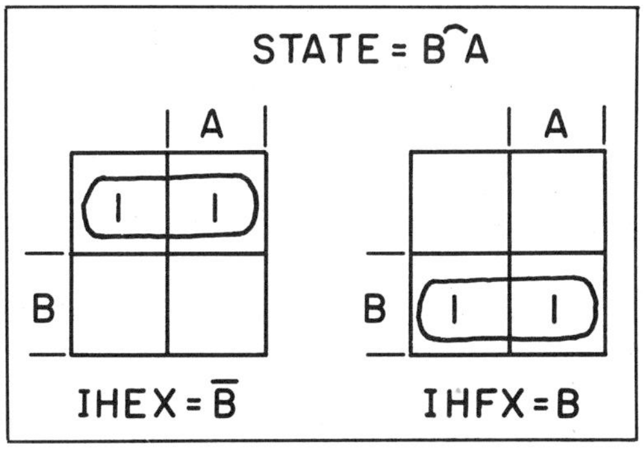

Figure 4.30 The Output Functions for the JK Flip-Flop.

All the link paths leading to the next state of 1 for variable A are highlighted in Figure 4.31. The conditions for each of these paths are entered into the next-state map for A shown in Figure 4.32. In state (a) enter YC·YJ. In state (b) enter 1, since the sum of both path conditions is 1. In state (c) enter $\overline{YC\cdot YK}$, since the path is the 0 exit of the condition box. In state (d) enter 0, since all paths lead to 0 for the next state of variable A.

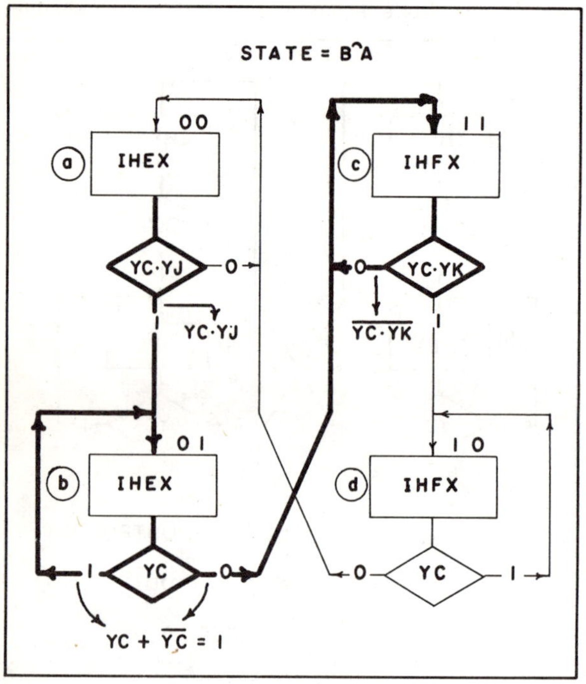

Figure 4.31 All the Link Paths Leading to the Next State of A = 1 Are Darkened. The Conditions for These Paths Are Entered Into the Next-State Map for Variable A.

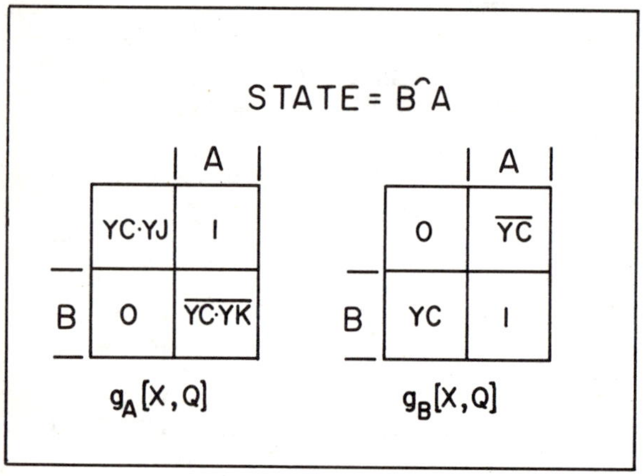

Figure 4.32 The Next State Maps for Variables A and B in the JK Flip-Flop.

All the link paths leading to the next state of 1 for variable B are highlighted in Figure 4.33. The condition for each of these paths are entered into the next-state map of B shown in Figure 4.32. In state (a) enter 0, since neither link path leads to next state of 1 for B. In state (b) enter $\overline{YC}$, since the link path from (b) exists on the 0 of the condition box for the next state of 1 for B. In state (c) enter 1, since all link paths lead to 1 for the next state of B; and finally, enter YC in state (d).

The two next-state maps are read using the rules of map-entered variables. First, circle all the 1's counting the map-entered variables as 0's. Second, circle the map-entered variables one at a time counting all 1's as don't cares and all other unlike variables as 0's. The results of the encirclements and map reading are shown in Figure 4.34. These two equations form the next-state function for the JK flip-flop.

Normal Boolean algebra is used to manipulate the next-state equations into a form directly implementable in gates, after the delay operator is replaced by an equal sign as was done in the RS example. Suppose that NAND gates are available. The two equations can be worked into the following forms

$$A = \overline{\overline{A\cdot(\overline{B\cdot YC\cdot YK})}\cdot\overline{\overline{B}\cdot YC\cdot YJ}}$$

$$B = \overline{\overline{\overline{YC}\cdot A}\cdot\overline{B\cdot\overline{\overline{A}\cdot\overline{YC}}}}$$

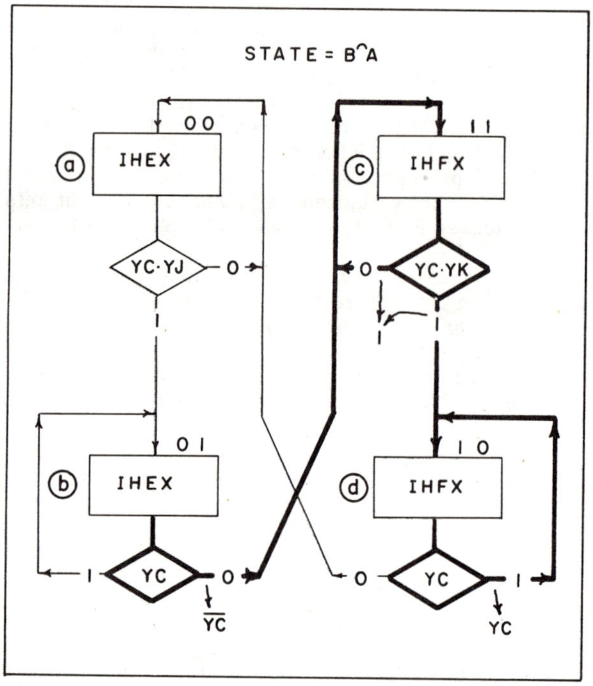

Figure 4.33 All the Link Paths Leading to the Next State of B = 1 Are Darkened.

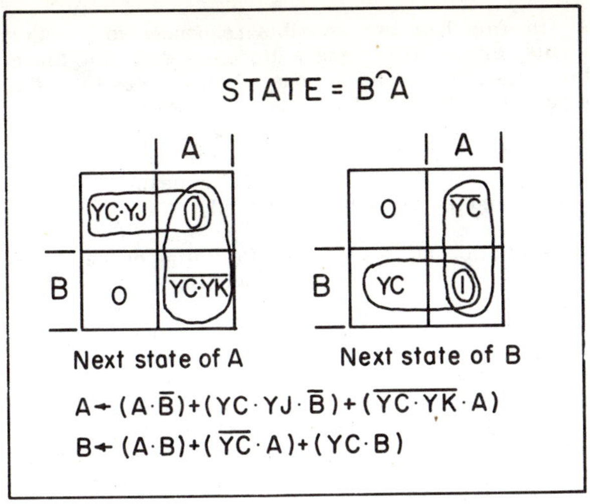

Figure 4.34 The Map Reading of the Next State Functions for the State Variables in the JK Flip-Flop.

To draw the gates for these equations, assume the existence of A and B as two points on the paper. Work from these two points to form $\bar{A}$ and $\bar{B}$, and then, using these terms and the inputs, form all the gates required to generate A and B, which were assumed at the start. Finally, add the required gating for the output function, which in this example amounts to a connection to B and to $\bar{B}$. The final combined circuit for both partial next-state functions and the output function yields a design as shown in Figure 4.35. Note that this design was a direct implementation of the next-state function and no attention was paid to hazzards, which will be covered in Section 4.9. Each partial next-state function of an asynchronous machine, such as this one, must be free of hazzards. It turns out that this design is free of such problems; therefore it is finished. This example has demonstrated the procedure of going directly from the ASM chart to maps, to equations and then to gates using the next-state function and the output function.

The JK flip-flop can be made into the D or the T flip flop by minor changes in the JK inputs. As was discussed, the D flip-flop has an ASM chart where $J = D$ and $K = \bar{D}$, and the T flip-flop has an ASM chart where $T = J = K$. The symbols for the JK, D and T flip-flops are summarized in Figure 4.36.

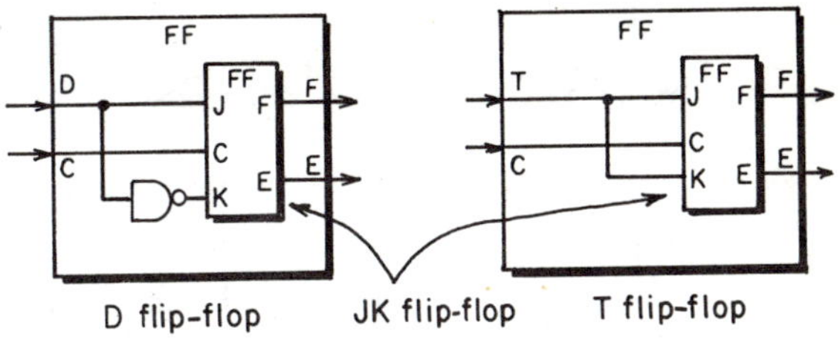

Figure 4.36 The JK, D and T Symbols.

Before leaving the subject of designing flip-flops, consider the design of a flip-flop which is not a one's catcher. Such a flip-flop could recover when the input changes back to 0 before the clock. Figure 4.37 shows the ASM chart for a flip-flop with this feature. Notice that the J and K inputs can return the flip-flop to states ⓐ or ⓒ after the flip-flop reaches states ⓑ or ⓓ. Sufficient time should be allowed for the state to settle before the clock goes to 0. This time depends on the internal gate delays and not the clock. This flip-flop is called edge-triggered. The design of this flip-flop follows the same procedure as used in the last example, but the resulting gate circuit is more complex. A suggested map uses YC, YK, A, and B as the map variables and leaves YJ as a map-entered variable. The details of this design are left as an exercise.

Another useful logic element is a GATE LATCH. This circuit can be confused with a clocked D flip-flop because it behaves somewhat the same for storing data but differs in that the data passes through when the gate input is 1. Figure 4.38 shows the ASM chart for this device and points out there are only two states as in an RS flip-flop. Neither the RS flip-flop nor the gated latch can be used alone as delay elements in synchronous machines; although the gated latch is often used as a temporary memory in logic systems. The design of this machine is left as an exercise.

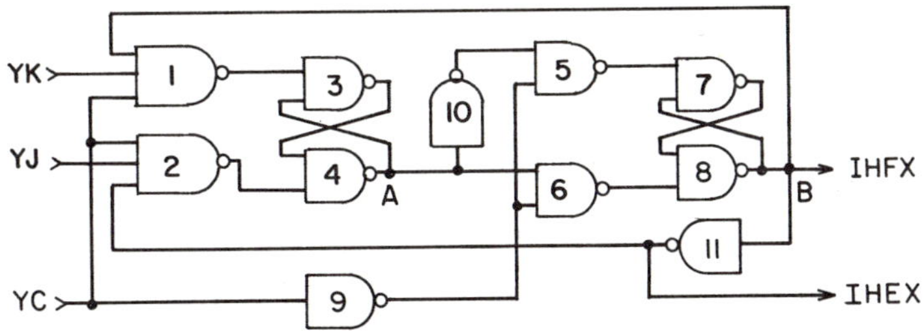

Figure 4.35 A NAND Gate Implementation of the Next-State Equation for a JK Flip-Flop. Gates 10 and 11 are eliminated in the RS Implementation of the JK Flip-Flop in Figure 4.42.

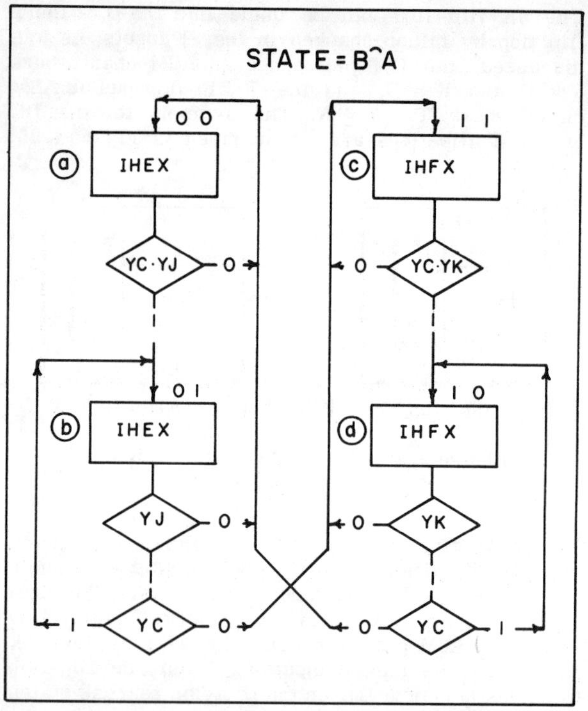

Figure 4.37 The ASM Chart of an Edge Triggered JK Flip-Flop.

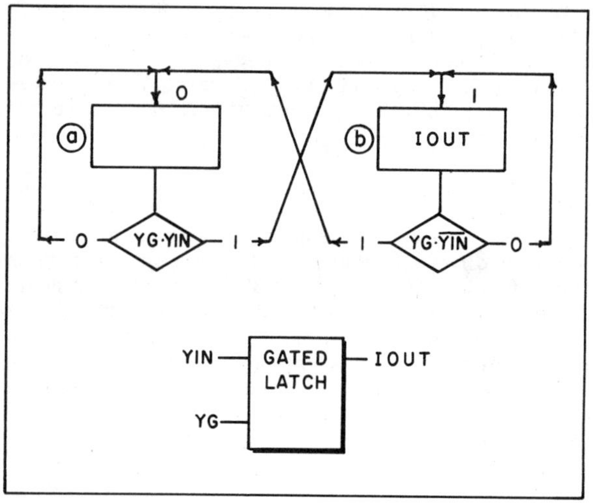

Figure 4.38 The Description of a Gated Latch

Use of the Excitation Table for Flip Flop Inputs

It would be possible to continue designing state machines as just described; however, the design is easier if flip-flops are used for the state variables. Flip-flops are themselves large building blocks replacing many gates. Flip-flops are incorporated in the design by translating the next-state information into flip-flop input equations by using an EXCITATION TABLE.

A flip-flop has two possible responses to a set of inputs, either it changes state or it does not. Since the flip-flop can be in one of two states when the inputs are applied, there are four possible TRANSITIONS described by an arrow between the two states, i.e., 0→0, 0→1, 1→0, and 1→1. (The arrow used to describe transitions is different in meaning than the arrow used for the delay operator.)

An excitation table consists of a listing of the inputs required to produce a given transition in a flip-flop. One way to generate such a list is to read the ASM chart description of the flip-flop. Figure 4.39 gives the excitation table for an RS flip-flop. Notice the inputs required for the four basic transitions include some don't care entries. There are two other transitions entries which are used to directly form the input equations from the ASM chart whenever a conditional transition produces two different variable transitions.

The excitation table is used to translate the ASM chart description of a state machine into maps for the inputs of the state variable flip-flops. Each state should have a state assignment, and each state variable is synthesized with a flip-flop. For example, a design using three state variables and being implemented with flip-flops would require six maps for the flip-flops inputs. The procedure for filling in the maps is similar to generating the next-state equations. First, fill in all unused states with don't cares. Second, identify the type of transition for each variable in each state of the ASM chart, and enter the corresponding value from the excitation table into the correct map. When the transition depends upon a qualifier, use map-entered variables to enter the conditions specified in the excitation table. The resulting maps can be read to yield the input equations for the state-variable flip-flops.

Transition	Inputs R	Inputs S
0 → 0	-	0
0 → 1	0	1
1 → 0	1	0
1 → 1	0	-
Conditional 0 → 0 or 0 → 1	0	Cond. for 0 → 1
Conditional 1 → 0 or 1 → 1	Cond. for 1 → 0	0

Figure 4.39 The Excitation Table for an RS Flip-Flop. "Cond." Stands for "Condition".

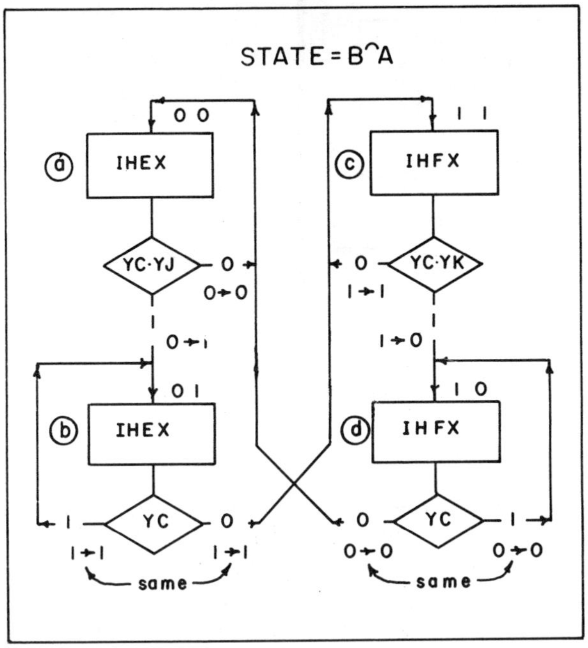

Figure 4.40 The JK Flip-Flop With All Transitions Indicated for State Variable A.

To illustrate the ease with which the input equations are derived, the JK flip-flop, Figure 4.29, will be designed using RS flip-flops for the state variables. Since all four states are used, each map square must be filled from the excitation table. Figure 4.40 illustrates how the transitions are identified for variable A. In state (a) variable A has a transition which is either 0→0 or 0→1 depending on the condition YC·YJ. According to the excitation table, such a transition has the R input set to 0 and the S input set to the condition for the 0→1 transition, which is YC·YJ. These entries are seen in Figure 4.41, state (a) of maps RA and SA respectively. The remaining map squares are filled in a similar manner. The equations are seen below each map as read using map-entered variables.

From the earlier discussion of the RS flip-flop, a very simple implementation using two NAND gates had negated inputs $\overline{YS}$ and $\overline{YR}$. The input equations for A and B can be changed to this form to produce

$$\overline{SA} = \overline{YC \cdot YJ \cdot \overline{B}}$$

$$\overline{RA} = \overline{YC \cdot YK \cdot B}$$

$$\overline{SB} = \overline{\overline{YC} \cdot A}$$

$$\overline{RB} = \overline{\overline{YC} \cdot \overline{A}} \ .$$

These equations can be combined with the RS flip flops to produce the logic circuit in Figure 4.42. The two flip-flops are sometimes called the MASTER and SLAVE, as indicated.

When the JK flip-flop using RS flip-flops, in Figure 4.42, is compared with the JK flip-flop using no RS flip-flops, in Figure 4.35, very little difference is seen. The only real difference is the absence of gate 10 and 11 in the RS version. These two gates are apparently eliminated by the assumption made in using the RS flip-flops, which is that R and S will never occur at the same time. As long as this assumption is correct, the RS version will work the same as the first version of the JK flip-flop.

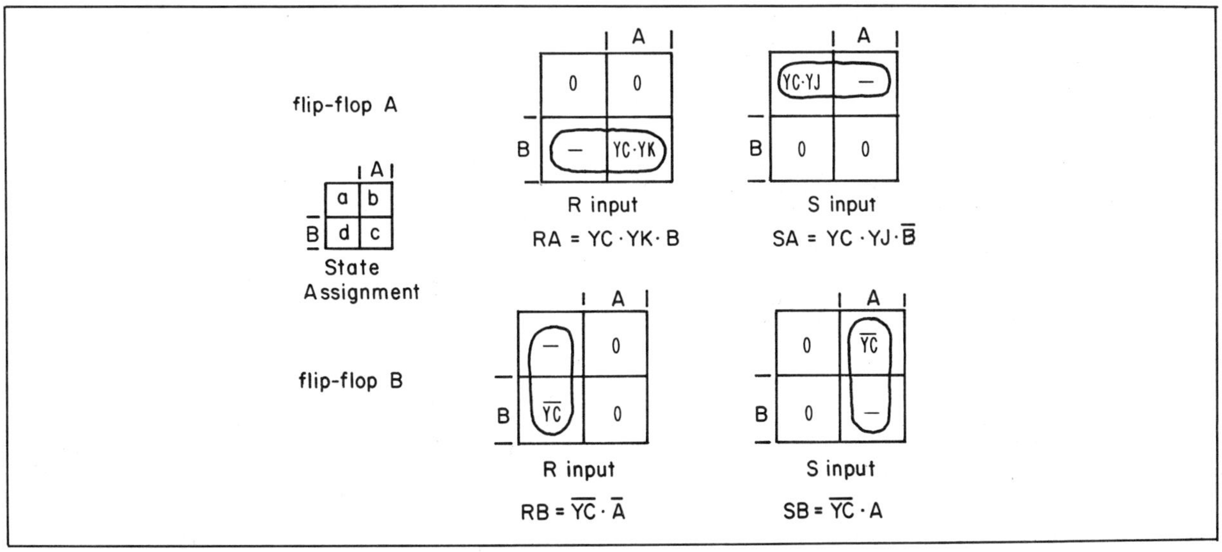

Figure 4.41 The Input Maps for RS Flip-Flops to Synthesize the JK Flip-Flop.

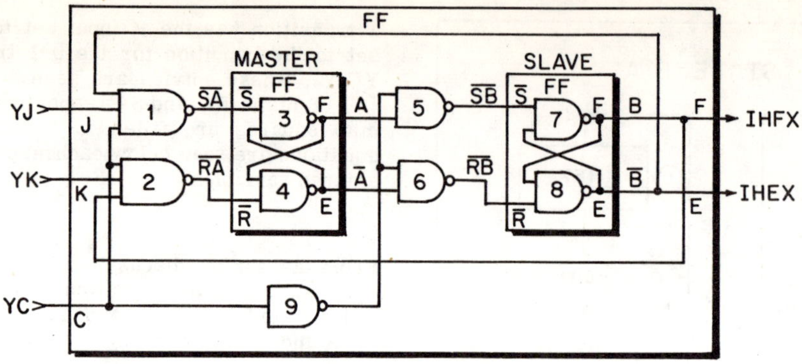

Figure 4.42 A NAND Gate Implementation of the Next-State Equations for a JK Flip-Flop Using RS Flip-Flops. (Also known as a Master-Slave Flip-Flop)

The D, T and JK flip-flops can be also used as state memory elements. Figure 4.43 gives the excitation table for these three flip-flops.

Transition	Flip-Flop Inputs			
	D	T	J	K
0 → 0 0 → 1 1 → 0 1 → 1	0 1 0 1	0 1 1 0	0 1 - -	- - 1 0
Conditional 1 → 0 or 1 → 1	Cond. for 1 → 1	Cond. for 1 → 0	-	Cond. for 1 → 0
Conditional 0 → 1 or 0 → 0	Cond. for 0 → 1	Cond. for 0 → 1	Cond. for 0 → 1	-

Figure 4.43 The Excitation Table for the D, T and JK Flip-Flops. "Cond." stands for "Condition".

The advantage of using a D, T, or JK flip-flop is that the state time can be defined by an external reference signal, usually called the CLOCK.

Figure 4.43 reveals some simplifications for developing the input maps for these flip-flops, which are as follows:

1. The D flip-flop-input equals the next state of the flip flop.

2. The T flip-flop input equals 1 for any change in the flip-flop state and is 0 otherwise.

3. The J input is don't care when the flip-flop = 1, and the K input is don't care when the flip-flop = 0. A 1 is entered in the remaining states only when there is a change in the flip-flop state.

In Chapter 3, a decade counter was described as an example of a class 2, direct transition, state machine. It will serve as a good comparison for an implementation with D, T, and JK flip-flops. To achieve the comparison, the four input equations are developed side by side for each flip-flop serving as a state variable of the counter. Figure 4.44 gives a tabular description of the decade counter. It then lists the maps for the D, T, J, and K inputs determined by using the simplified rules above.

Each set of maps provides a solution for the decade counter, using one of the flip-flop types and additional gates. The input equations, read from the maps, are summarized as follows:

$DA = \overline{A}$ $\quad TA = 1$

$DB = (\overline{D} \cdot A \cdot \overline{B}) + (B \cdot \overline{A})$ $\quad TB = A \cdot \overline{D}$

$DC = (\overline{B} \cdot C) + (\overline{A} \cdot C) + (\overline{C} \cdot A \cdot B)$ $\quad TC = A \cdot B$

$DD = (\overline{A} \cdot D) + (A \cdot B \cdot C)$ $\quad TD = (A \cdot D) + (A \cdot B \cdot C)$

$JA = 1$ $\quad KA = 1$

$JB = A \cdot \overline{D}$ $\quad KB = A$

$JC = A \cdot B$ $\quad KC = A \cdot B$

$JD = A \cdot B \cdot C$ $\quad KD = A$

From these equations, it is clear that the D input equations are the most complex and the J and K the simplest. This difference is also evident in the gates required to synthesize the counter from these equations. Figure 4.45 shows the D flip-flop counter implemented with AND-OR gates, Figure 4.46 shows the T flip-flop counter implemented with NAND-NOR gates, and Figure 4.47 shows the JK counter implemented with AND gates.

NAME	X	g[X]
	D⌒C⌒B⌒A	D⌒C⌒B⌒A
0	0 0 0 0	0 0 0 1
1	0 0 0 1	0 0 1 0
2	0 0 1 0	0 0 1 1
3	0 0 1 1	0 1 0 0
4	0 1 0 0	0 1 0 1
5	0 1 0 1	0 1 1 0
6	0 1 1 0	0 1 1 1
7	0 1 1 1	1 0 0 0
8	1 0 0 0	1 0 0 1
9	1 0 0 1	0 0 0 0

NEXT-STATE TABLE

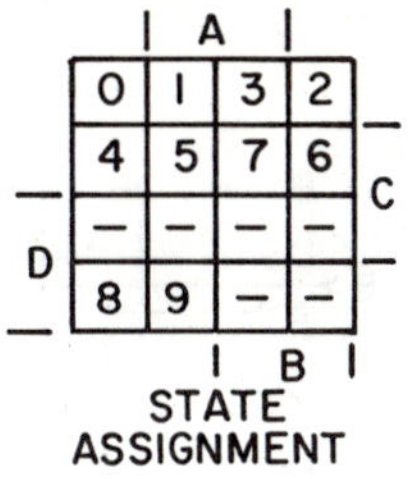

all maps below have the same assignment

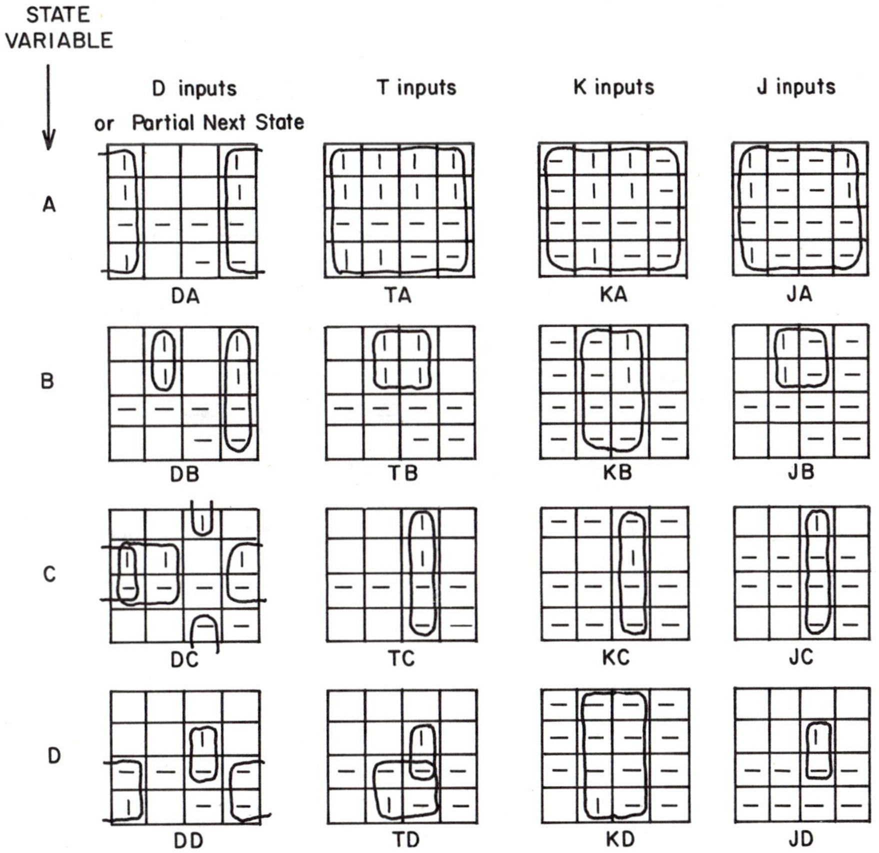

Figure 4.44 D, T, and JK Input Maps for a Decade Counter.

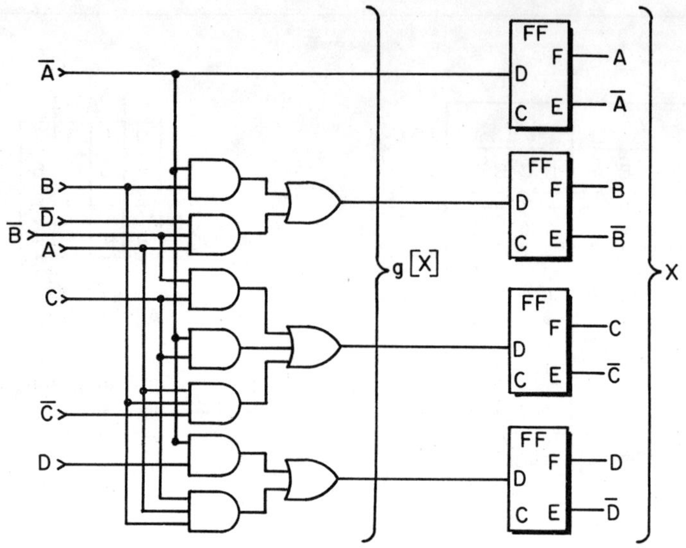

Figure 4.45 A Synchronous Decade Counter Using Delay Flip-Flops.

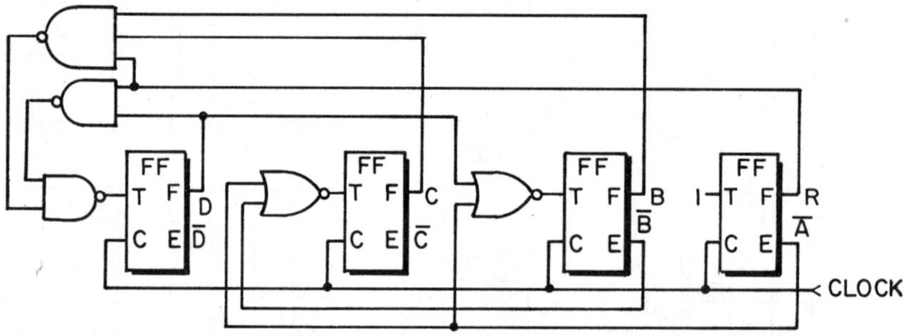

Figure 4.46 A Decade Counter Implemented with T Flip-Flops.

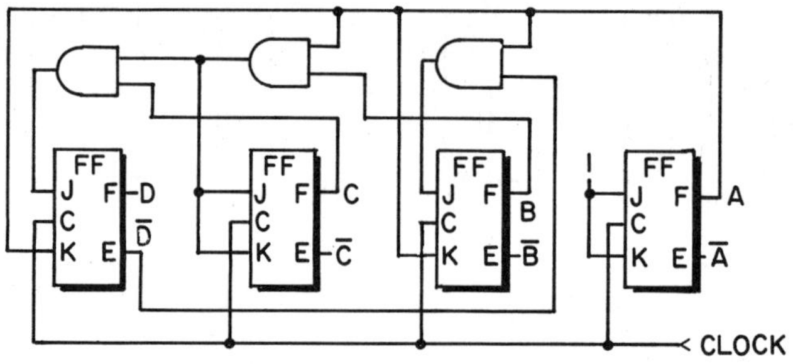

Figure 4.47 A Synchronous Decade Counter Using JK Flip-Flops.

While counters are fine for comparative implementations, a more interesting problem is converting an ASM chart directly to JK flip-flops. Figure 4.48 describes a simple 3-state machine to be used as an example. The two state variables, A and B, are to be implemented using JK flip-flops. For illustration, each step will be explained. First, draw a state assignment map as shown in Figure 4.49(a). There is one unused state which is called don't care. Second, draw a map for each flip-flop input, in this example four maps, and fill in the unused states with don't cares, Figure 4.49(b). Third, from the observation of the JK flip-flop excitation table that the J input is don't care when the flip-flop is 0, put don't cares in squares where a variable A is 0 in KA, a 1 in JA; and where variable B is a 0 in KB and a 1 in JB, Figure 4.49(c). Fourth, fill in the remaining squares by reading the ASM chart to find what conditions cause a change in the state variable being considered. In the KA map, state (b) is filled with a 1 by observing that the A variable always changes when leaving state (b). In the JA map, state (a) is filled with YPON because the A variable changes only when YPON=1. In the state map, state (c). is filled with 0 because the A variable does not change when leaving state (c). In the KB map, state (c) is filled with a 1 because the B variable always changes when leaving state (c). In the JB map, state (b) is filled with $\overline{NSNK}$ because the B variable changes only when NSNK=0. Finally, in the same map state (a) is filled with 0 because the B variable does not change when leaving state (a). The maps can be read as shown in Figure 4.49(d). The final circuit diagram is shown in Figure 4.50 along with the simple output maps and gate reductions. This simple procedure can be repeated for more complex designs to find the input equations for JK flip-flops to implement the next-state function. A more complex example is worked out in Section 4.10.

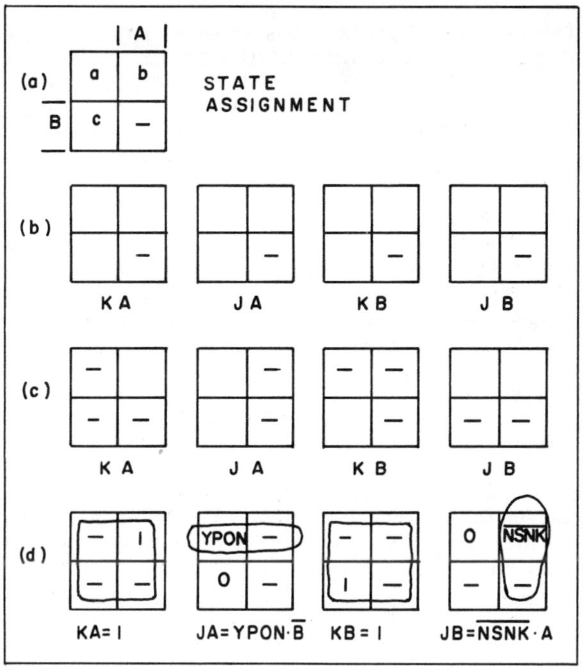

Figure 4.49 Successive Solution Steps for the 3-State Machine.

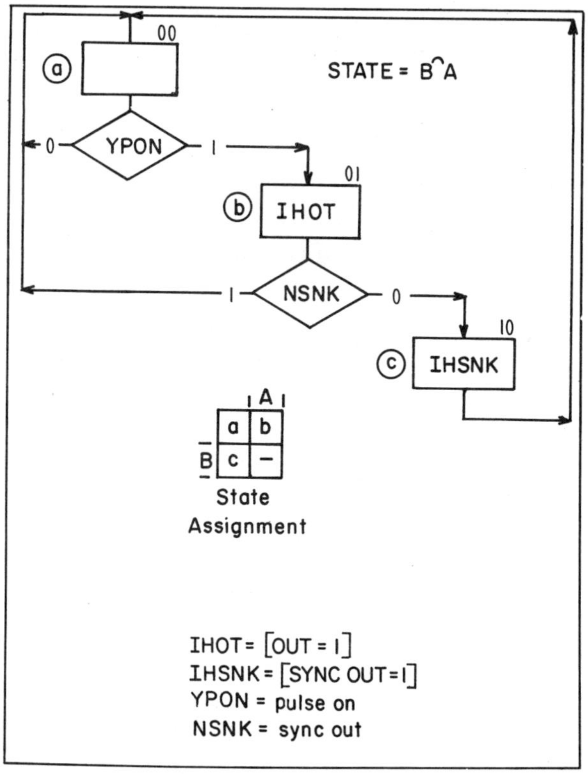

Figure 4.48 Example of a 3-State Conditional Transition ASM Chart.

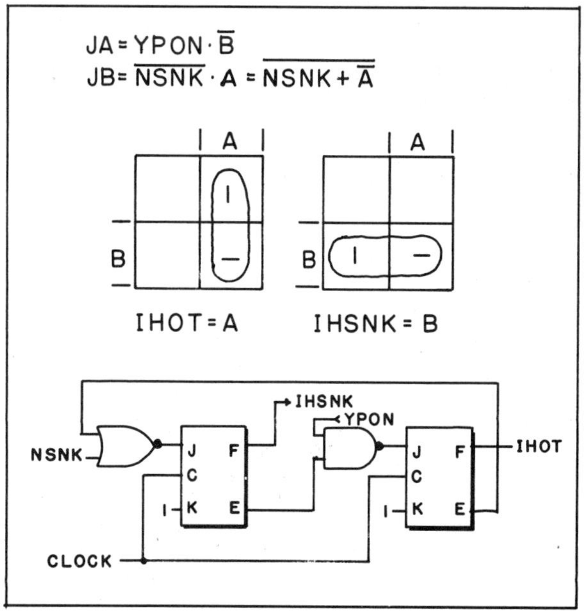

Figure 4.50 The Final Gate Reduction and Solution for the 3-State Machine.

4.8 STATE ASSIGNMENT

This section describes the factors in a state machine implementation that affect the choice of state codes. The concepts covered are MINIMUM STATE LOCUS, REDUCED DEPENDENCY, TRANSITION RACES, and OUTPUT RACES.

The process of giving codes to each state of a state machine is called ASSIGNING STATES. The set of codes is called the STATE ASSIGNMENT. In the state machines described previously in this chapter, the state assignments were given and the synthesis process proceeded from this point. Thus, state assignment is one of the first steps in synthesis.

The discussion of state assignment has been reserved until last because it depends upon understanding the rest of the synthesis process. For example, the choice of states greatly affects the gate complexity of both the output function and the next-state function; and the details of choosing states to reduce complexity involves knowing how these two functions are formed. However, the complexity of the gating does not affect whether the state machine works or not. There are many workable assignments.

State assignment can be approached in two ways. One way is to consider the entire machine, including the possibilities of races, and find the one best assignment that requires the fewest number of gates for fault-free operation. Although a great deal of work has been done to solve this general problem, the indications are that the general problem is too complex to consider solving without a lengthy computer search. Another way to approach state assignment is to consider each state separately, minimizing the probable contribution to the gate complexity made by each state. This problem is far easier than the general one and, from experience, proves to be extremely practical while yielding optimum or near-optimum results. In this section, the state-by-state approach will be considered, the resultant solution being called a REASONABLE MINIMUM.

Minimum State Locus

The concept of minimum state locus is based on minimizing the probable contribution to the gate complexity of the next-state function on a state-by-state basis. Basically, the complexity of gating in the next-state function depends on the number of state variables involved in the transition from one state to the next. The number of variables involved in a transition is called the BIT DISTANCE between the two states. For every ASM block there is a bit distance associated with every exit path. The sum of all the bit distances of all the exit paths in a state machine is called the STATE LOCUS. The state locus is dependent upon the state assignment as well as the algorithm. The probable gate complexity is reduced for a state assignment with a smaller state locus.

The state locus is only an approximation which produces the most useful results in the most complicated link structures where other combining factors are less likely to be effective. A decade counter demonstrates how the state locus can be misleading in a simple problem. Figure 4.51 gives two state assignments for a 10-state direct transition machine. The first assignment is based on a single bit change or a UNIT DISTANCE between successive states and is called a GRAY CODE. The gray code has the minimum state locus for any even number of states in a counter chain, such as in this example. A gray code cannot be found for an odd number of states in a chain. The state locus for the gray code counter is 10, one bit change for each count. The other assignment is based on a binary weighted coding of the bits to represent the count. The state locus for this assignment is 18.

STATE = D C B A

GRAY CODE ASSIGNMENT	BINARY WEIGHTED ASSIGNMENT
0000	0000
0001	0001
0011	0010
0010	0011
0110	0100
0111	0101
0101	0110
0100	0111
1100	1000
1000	1001
STATE LOCUS = 10	STATE LOCUS = 18

Figure 4.51 Two Assignments for a 10-State Counter.

The relative complexity of the next-state functions for these two assignments is seen by comparing the partial next-state equations for the state variables, as follows:

g [X] with the GRAY CODE ASSIGNMENT (16 literal terms):

$$g_A[X] = (\overline{B} \cdot \overline{C} \cdot \overline{D}) + (C \cdot B)$$
$$g_B[X] = (\overline{C} \cdot A) + (\overline{A} \cdot B)$$
$$g_C[X] = (C \cdot \overline{D}) + (\overline{A} \cdot B)$$
$$g_D[X] = C \cdot \overline{A} \cdot \overline{B}$$

g [X] with the WEIGHTED BINARY ASSIGNMENT (18 literal terms):

$$g_A[X] = \overline{A}$$
$$g_B[X] = (\overline{D} \cdot A \cdot \overline{B}) + (B \cdot \overline{A})$$
$$g_C[X] = (\overline{B} \cdot C) + (A \cdot C) + (\overline{C} \cdot A \cdot B)$$
$$g_D[X] = (\overline{A} \cdot D) + (A \cdot B \cdot C)$$

As suspected, the unit distance code with the smaller state locus is slightly simpler than the binary code. The surprise is that when implemented with JK flip-flops, the binary code solution is simpler, which can be seen by the following comparison of the flip-flop input equations:

GRAY CODE

$$JA = KA = \overline{D}((B \cdot \overline{C}) + (C \cdot \overline{B}))$$
$$JB = A \cdot \overline{C} \qquad KB = A \cdot C$$
$$JC = \overline{A} \cdot B \qquad KC = D$$
$$JD = \overline{A} \cdot \overline{B} \cdot C \qquad KC = \overline{A}$$

BINARY CODE

$$JA = 1 \qquad KA = 1$$
$$JB = A \cdot \overline{D} \qquad KB = A$$
$$JC = A \cdot B \qquad KC = AB$$
$$JD = A \cdot B \cdot C \qquad KD = A$$

In this simple problem, the particular binary assignment produces a simpler JK implementation because there is a JK-like symmetry in the count sequence. In more complex problems, with conditional branching and asymmetric structures typical of most algorithms, the state locus is a more reliable indicator. However, the state locus does not include the contribution to the next-state function by the inputs. These effects are covered next by reduced dependency.

Reduced Dependency

A state assignment for a conditional transition made according to the concept of REDUCED DEPENDENCY makes a smaller probable contribution to the next-state function gating than an assignment by minimum state locus. Figure 4.52 illustrates a sample conditional transition with two proposed assignments. The first assignment is based on the minimum state locus, that is, one bit change from state (ℓ) to state (m) and one bit change from state (ℓ) to state (n). For this assignment, the partial next-state equations are:

$$\left\{\begin{matrix} g[\ell, Q] \\ \text{with STATE} = B \frown A \end{matrix}\right\} = \begin{bmatrix} g_A\,[\ell, Q] = \overline{YQ} \\ g_B\,[\ell, Q] = YQ \end{bmatrix}$$

and both variables depend on YQ.

For the second assignment, the partial next-state equations are:

$$g[\ell, Q] = \begin{bmatrix} g_A\,[\ell, Q] = 1 \\ g_B\,[\ell, Q] = YQ \end{bmatrix} \text{(changes for both transitions)}$$

and only the B variable depends on YQ. The A variable always changes, saving a gate input. Thus, the second assignment is called a reduced dependency assignment, compared to the minimum state locus assignment. A reduced dependency assignment saves gate inputs in the next-state function and therefore is an improvement on the minimum state locus approach. The reduced dependency concept can be understood for more complex problems by considering the transitions between states as CHANGE VECTORS. The transition from (ℓ) to (m) and (ℓ) to (n) could then be represented in two ways. Using a minimum state locus assignment, the transition would be represented by Figure 4.53,

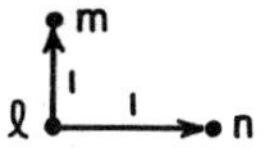

Figure 4.53 Minimum State Locus Change Vectors.

indicating a one-bit change to (m) and (n) by an arrow for each vector. The reduced dependency assignment would be represented by Figure 4.54. From this

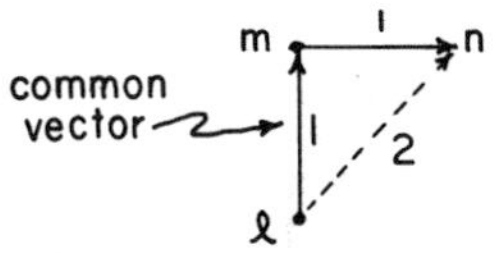

Figure 4.54 Reduced Dependency Change Vectors.

diagram, the change vector (ℓ) to (m) is common to both transitions. The dotted line indicates the two-variable transition from (ℓ) to (n), which is really the sum of two single-variable change vectors. A state with five next states labeled (a), (b), (c), (d), and (e) could also be given a reduced dependency assignment according to Figure 4.55. This assignment has many common change vectors to reduce the gate complexity. The dotted lines indicate the bit distance for each transition. From this discussion, finding a reduced dependency assignment is really equivalent to finding common change vectors, and finding common change vectors is aided by a new use of a map.

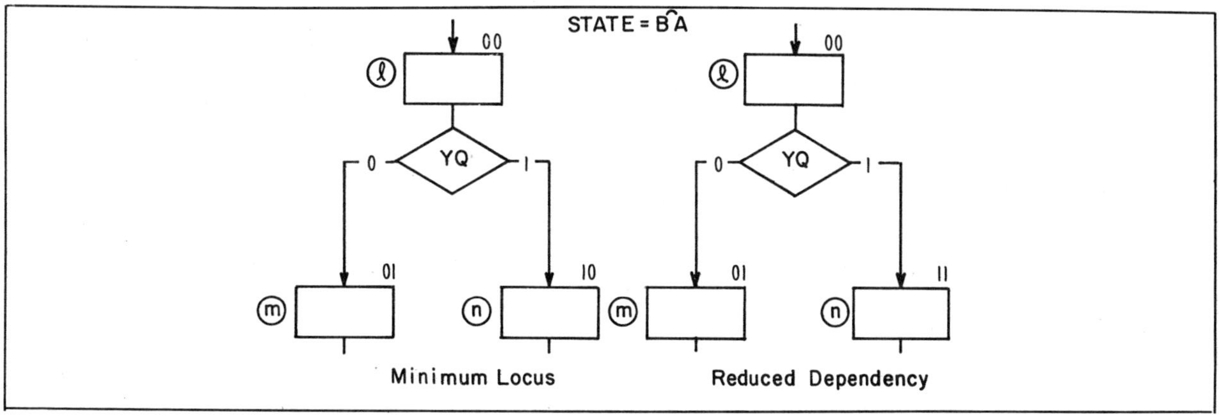

Figure 4.52 Two State Assignments.

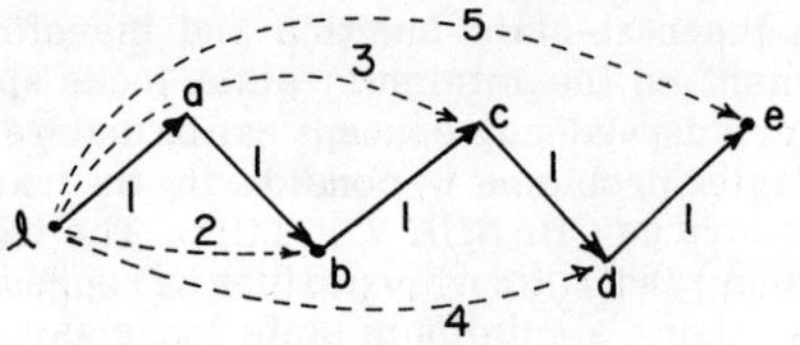

Figure 4.55 Reduced Dependency Change Vectors.

The choice of a reduced dependency state assignment for complex link paths is aided by a maplike diagram called a DEPENDENCY MAP. The diagram is made by drawing a map using all of the conditional variables as map variables and entering the next state resulting from each input combination. The map, then, describes the relationship between the next states and the conditional variables. The dependency map, drawn for the assignment problem in Figure 4.52, is shown in Figure 4.56. From the map it can be seen that when YQ = 0, the next state is ⓜ; and when YQ = 1, the next state is ⓝ. Thus, the map describes the next-state dependency for state ⓛ.

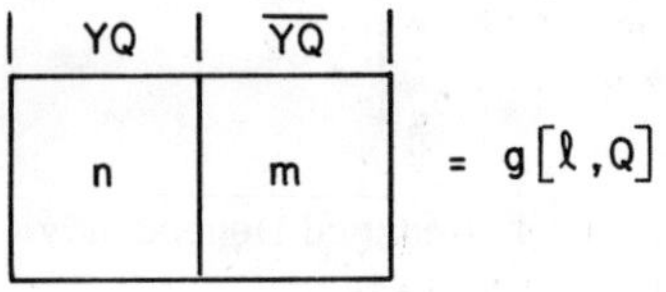

Figure 4.56 A Dependency Map.

The dependency map is used to find equations for the change vectors. Each change vector is indicated on the dependency map by encirclements of the common next states for the change vector. Each change vector points to a specific next state, which is used as the name of the change vector. The equation for the change vector is read from the corresponding dependency map encirclement. The function for the change vectors is called "v" for vector. The function, v, is a function of the state and the qualifiers, and thus has the form v[X,Q]. The change vector function is described by listing all the equations for the change vectors. Using

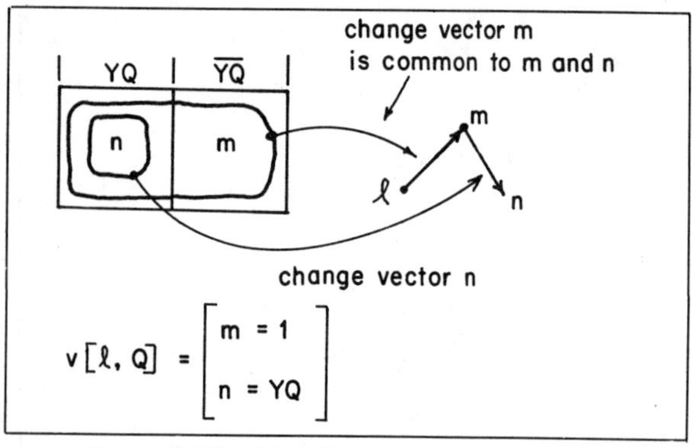

Figure 4.57 The Change Vector Function for Reduced Dependency.

the same dependency map as before, the change vector function for the reduced dependency assignment is obtained as shown in Figure 4.57. This dependency map may be compared to the dependency map for the minimum locus assignment, which is shown in Figure 4.58.

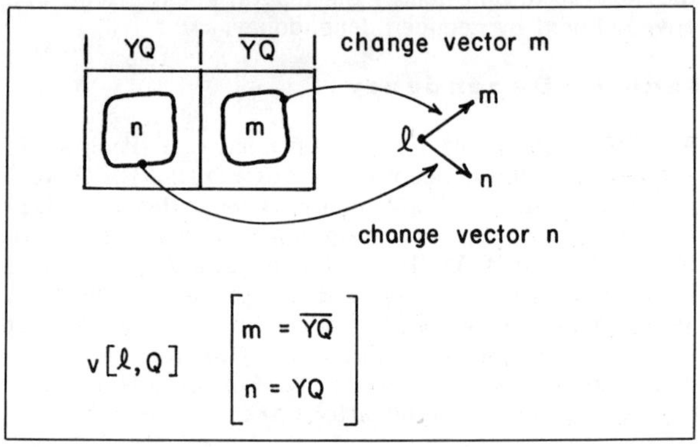

Figure 4.58 A Minimum State Locus Change Vector Equation.

A state assignment utilizing the common change vectors is made by assigning a particular state variable change to each change vector. The simplest assignment changes one state variable for each change vector. This assignment requires at least as many state variables as there are change vectors. In the previous example, the reduced dependency assignment was made by letting the transition of the A variable from 0 to 1 be the m transition vector, and the letting the transition of the B variable from 0 to 1 be the n transition of the B variable from 0 to 1 be the n transition transition vector, the partial next-state equation is the true or complement of the corresponding transition vector equation, depending on the next state of the state variable. The assignment for both minimum state locus and reduced dependency are summarized in Figure 4.59.

Figure 4.60 shows a more complex link structure for a state ⓛ going to state ⓜ, ⓝ, ⓟ, and ⓡ, along with the dependency map made by using YX, YY, and YZ as the map variables. A possible assignment is also given. In this state the dependency map transition vectors indicated may be compared with independent transition vectors as follows:

	m = YX		m = 1
v[ℓ, Q] =	n = $\overline{YX}$ · YY	v[ℓ, Q] =	n = $\overline{YX}$
independent vectors	r = $\overline{YX}$ · $\overline{YY}$ · $\overline{YZ}$	common vectors	r = $\overline{YX}$ · $\overline{YY}$
	p = $\overline{YX}$ · $\overline{YY}$ · YZ		p = $\overline{YX}$ · $\overline{YY}$ · YZ

Three terms are saved by the common change vectors.

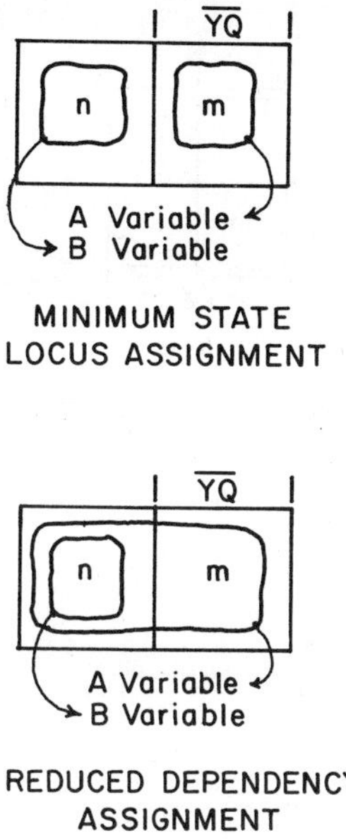

Figure 4.59 A Dependency Map Comparison for the Two Assignments of Variables to Change Vectors in Figure 4.52.

Sometimes a reduced dependency assignment using common change vectors involves more complex equations than some other assignment. For example, in Figure 4.61 the next state of ⓐ is itself, and thus no change vector is desired for some of the input conditions. Rather than listing the present state in the dependency map, a 0 is entered to indicate no change vector. The minimum encirclements for states ⓑ and ⓒ indicate that the change vectors are not common and that there is no simpler representation in finding common change vectors.

In a summary of a reduced dependency state assignment, the following rule applies:

> A state assignment using single-variable transition vectors, which are described by simple nested encirclements in the dependency map, is the most desirable reduced dependency assignment for the minimum probable contribution to the next-state function.

Transition Races

Although state locus and reduced dependency tend to simplify the next-state gating, some restrictions are imposed on the choice of states which may cause additional complexity. These restrictions are imposed to insure that the output function and the next-state function are error-free. An error in the output function is an output which occurs at unspecified times. An error in the next-state function is a transition to the wrong state. Both of these errors can be caused by the wrong choice of state vectors. The following section discusses the process of finding the right state vectors to avoid these problems.

The next-state function is based on the values of logic signals during the stable portion of the state time. When the state time is defined by the inputs, as in the implementation of the clocked JK flip-flop with gates, the stable period briefly follows the input until one of the state variables begins to change. This change may alter the outputs of the next-state gating in preparation for the next transition. The stable period, then, is defined only for the brief instant between the input transition and the change of the first variable. With this short time, only one variable can be depended upon to reliably assume the next state, and that variable is the one which detects the input first. This situation is called a RACE. Thus, in this type of state machine, the assignment of successive states in which two variables change may result in any of three next states having one or the other, or both, of the variables changed. This new state, in turn, may have a next-state change defined, causing a new transition, and so on until a state is finally reached for which the next state is defined as itself. This state is called the STABLE STATE. It is possible to account for all the intermediate states between stable states, but intermediate states are just wasteful where not required. The number of possible next states where the bit distance equals n is 2^n-1 states. If all states can be assigned unit distance codes, then there is only one next state. This assignment makes the best use of available states. The JK flip-flop has such an assignment.

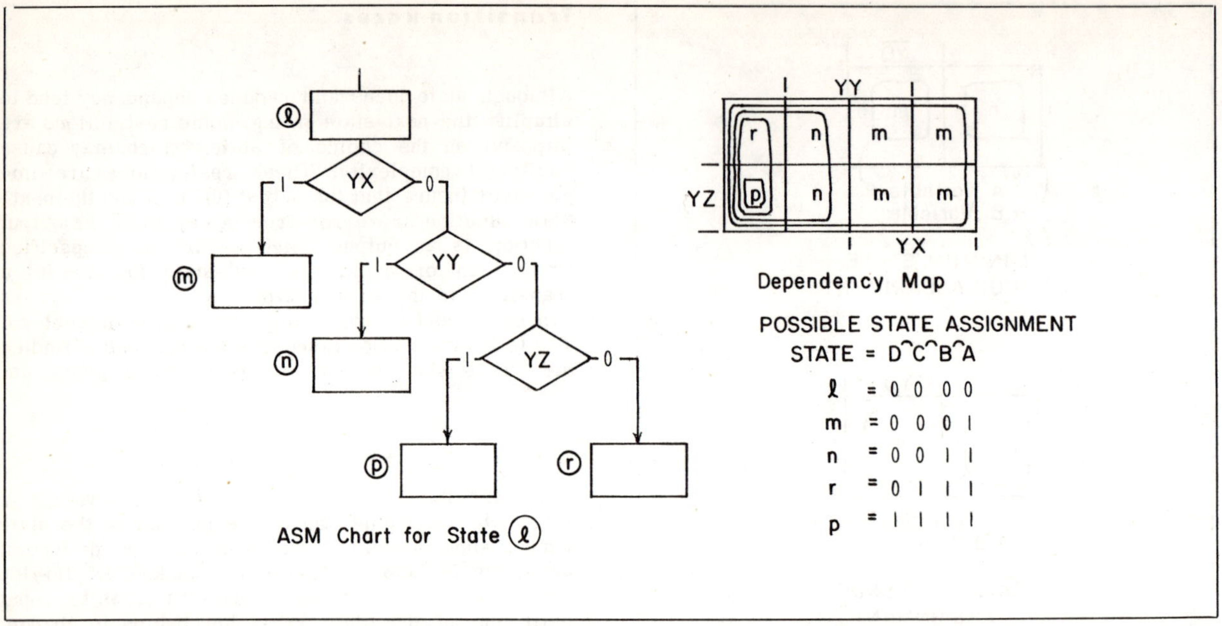

Figure 4.60 Using the Dependency Map to Determine a Reduced Dependency Assignment with a State with Four Next States.

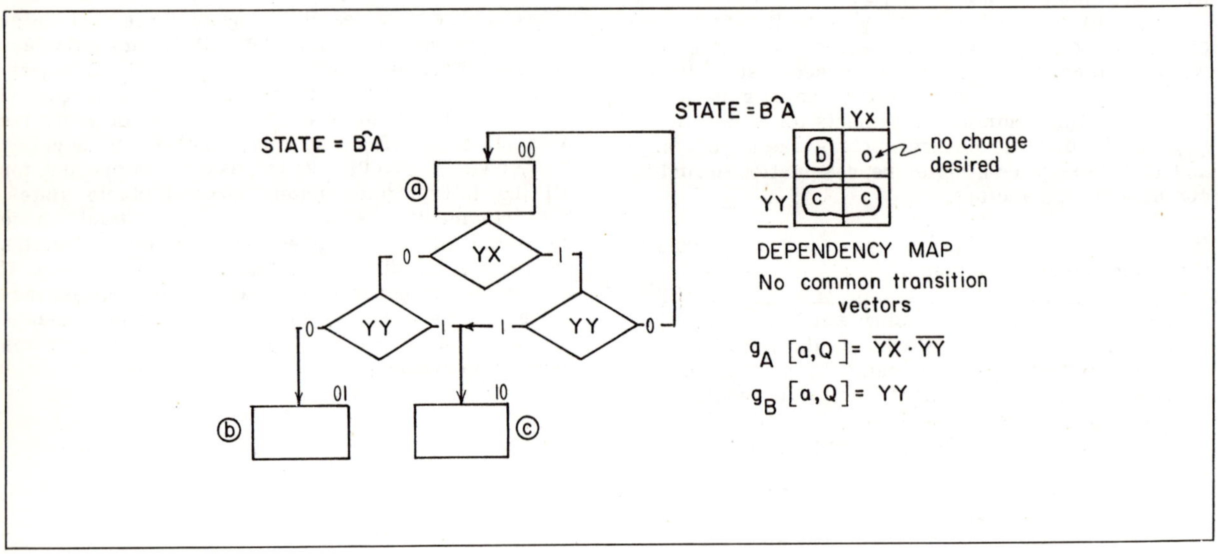

Figure 4.61 A State Linkage for Which Common Transition Vectors Would Complicate the Next-State Function.

The difficulty in assigning states to guarantee the correct transitions between stable states is partially overcome when the state variables are implemented with clocked flip-flops. Since the state time is defined by the clock, adequate time can be allowed for the gates generating the next-state function to reach a stable condition with stable inputs. Inputs which are stable during the stable part of the state time are called SYNCHRONOUS. Inputs which can change at any time independent of the state time are called ASYNCHRONOUS. Asynchronous inputs or qualifiers cause the same trouble in clocked machines as any input does in an unclocked machine. If more than one variable is dependent on the asynchronous qualifier, there is uncertainty in the next state. The following discussion assumes that any uncertainty in the next state is an error.

A criterion for a race-free state assignment in any state having asynchronous qualifiers can be stated as follows:

> To find a race-free state assignment, no more than one variable can depend on an asynchronous input for any combination of synchronous inputs in any state.

The variable dependency is represented by the next-state dependency map and partial next-state equations, as was discussed in finding a reduced dependency assignment. The best way to explore assigning states using the dependency map is to consider several examples. Figure 4.62 describes a state, ⓡ, with next states ⓐ, ⓑ, and ⓒ. There is one asynchronous qualifier, *YX, and one synchronous qualifier, YY. The star(*) signifies asynchronous qualifiers.

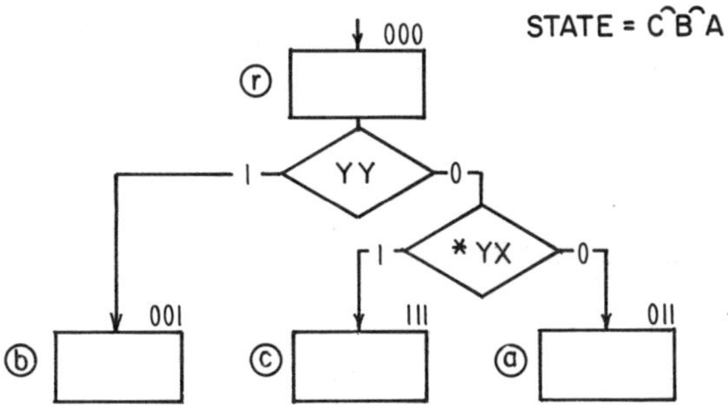

Figure 4.62 A State with Both Synchronous and Asynchronous Qualifiers which is given a Race-Free Reduced Dependency State Assignment.

The assignment given in Figure 4.62 was chosen to reduce dependency. The dependency map for the transition vectors and the partial next-state equations for this assignment are shown in Figure 4.63. Only the variable C is dependent on the asynchronous qualifier, *YX; therefore, the assignment is race-free.

Had a different selection for the state assignment been made, there might have been a race. Suppose that the transition vectors are chosen to minimize the state locus. The resulting dependency map and partial next-state equations would then be as shown in Figure 4.64. Variables B and C both now depend on *YX, so this assignment is not race-free.

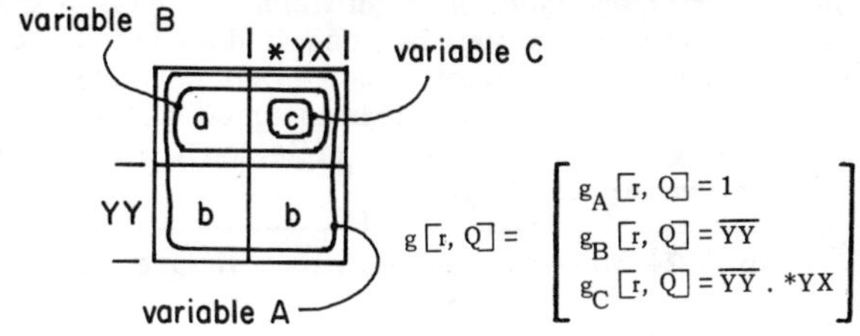

Figure 4.63 Reduced Dependency.

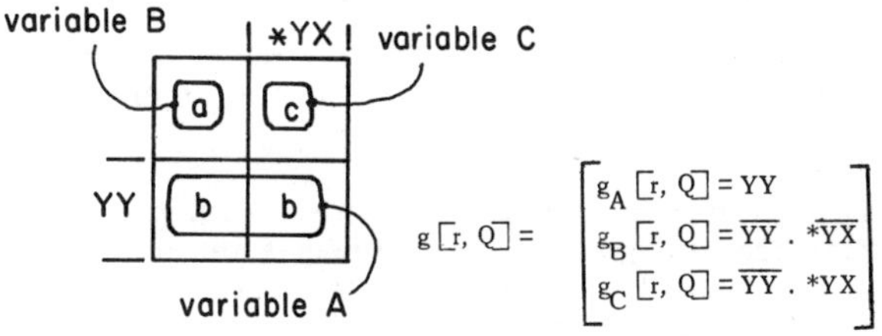

Figure 4.64 Minimum State Locus.

Figure 4.65 shows another example with one synchronous and one asynchronous qualifier on state ⓢ.

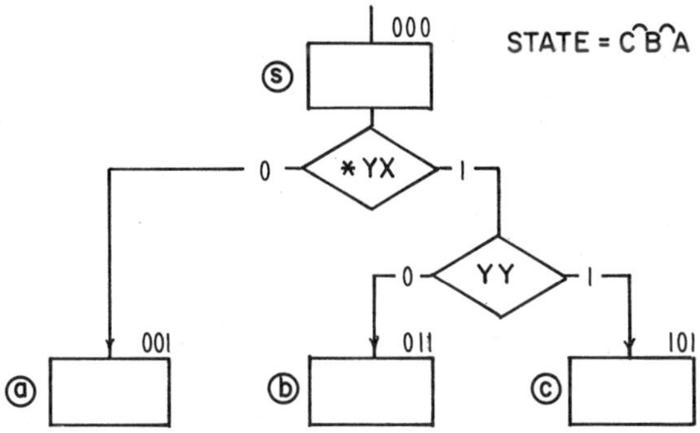

Figure 4.65 A State with a Race-Free State Assignment.

Assume the assignment for ⓐ, ⓑ, and ⓒ is unknown. The next-state dependency map for state ⓢ, assuming independent change vectors, and the partial next-state equations for a possible assignment, with one variable per change vector, are shown in Figure 4.66. This assignment is race prone because two variables depend on the asynchronous qualifier for a given synchronous input. If YY = 1, then variables A and C are dependent on *YX; and if YY = 0, then variables A and B are dependent on *YX. The reduced dependency assignment which solved the race in the previous problem is represented for this problem by Figure 4.67. This assignment is also race-prone because variables B and C both depend on *YX when YY = 1. There is a race-free assignment possible for this state, which is found as shown in Figure 4.68. With this assignment, either variable B or C depends on *YX for a given input YY, but never both. Thus, this assignment, also shown in Figure 4.59, is race-free.

The criterion for a race-free assignment can be expressed in terms of the change vector equations to eliminate the necessity of assigning variables and forming the partial next-state equations as was done in the previous example. It is sufficient to look at the change vector equations, because it does not matter which direction a variable changes; it only matters that it changes. If one variable is assumed to represent each change vector, then the criterion for a race-free assignment can be stated:

> An ASM state has a race-free assignment if only one of its associated partial change vector equations is dependent on an asynchronous qualifier for any combination of synchronous inputs.

This criterion serves as a quick check on a proposed ASM chart block to test for the possibility of a race-free assignment.

The partial next-state equations, developed for a race-free assignment, become map-entered variables in the formation of the complete next-state function. The process of state assignment has already done some of the synthesis work.

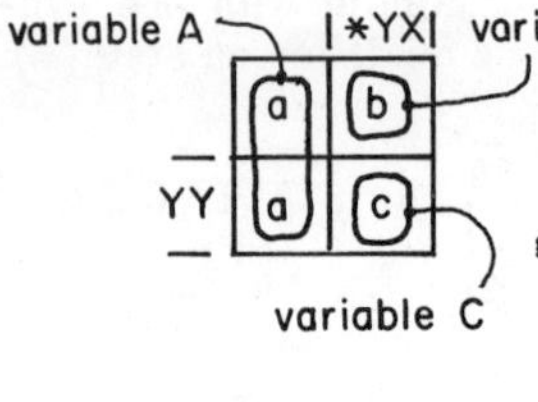

$$g[s,Q] = \begin{bmatrix} g_A[s,Q] = \overline{*YX} \\ g_B[s,Q] = \overline{YY} \cdot *YX \\ g_C[s,Q] = YY \cdot *YX \end{bmatrix}$$

Figure 4.66 Independent Change Vectors.

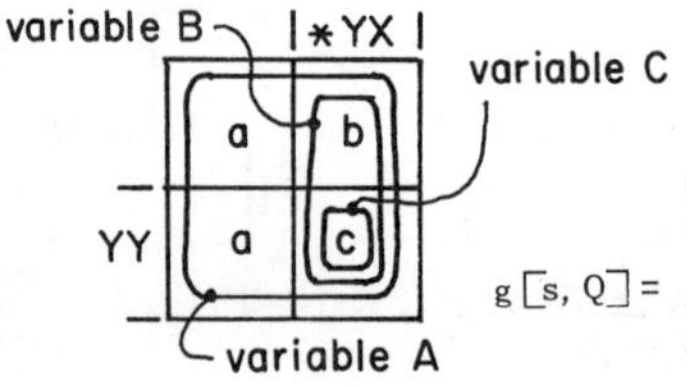

$$g[s,Q] = \begin{bmatrix} g_A[s,Q] = 1 \\ g_B[s,Q] = *YX \\ g_C[s,Q] = YY \cdot *YX \end{bmatrix}$$

Figure 4.67 Race Prone Vectors.

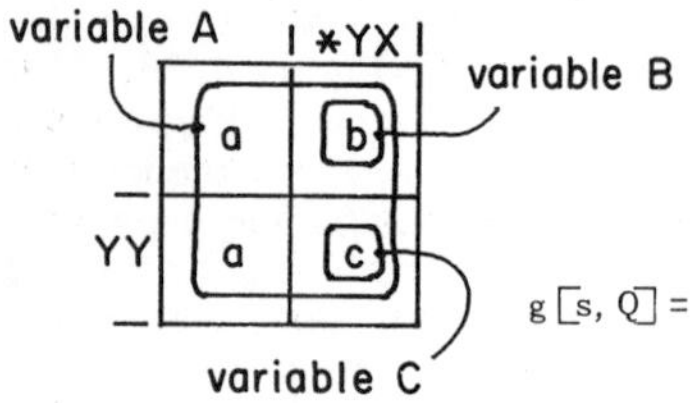

$$g[s,Q] = \begin{bmatrix} g_A[s,Q] = 1 \\ g_B[s,Q] = \overline{YY} \cdot *YX \\ g_C[s,Q] = YY \cdot *YX \end{bmatrix}$$

Figure 4.68 Race-Free Vectors.

Output Races

The last factor to affect the selection of a state assignment are the output races. Output races are generally caused by an output function which has more than one variable changing at a time and take the form of short pulses on outputs which are supposed to be stable. The pulses usually occur during the transition period when the synchronous variables are changing; they can be caused at other times by asynchronous inputs.

Output races during the transition period are significant only when the output is an immediate function, since an immediate function begins to respond as soon as any output occurs. The delay function ignores such races because it only responds to the stable portion of the state time. Both types of outputs may respond to races caused by asynchronous inputs, and for that reason asynchronous inputs should not be used to generate conditional outputs in a class 4 machine. The races remaining on immediate outputs occur during the transition period. When these races are intolerable for proper machine operation, they are called CRITICAL OUTPUT RACES. The process for determining output races and the means of avoiding the races centers on the output equations and the states involved in transitions. For every state transition which involves more than one state variable, the number of state codes possible during the transition period is 2^n, where n is the number of variables changing. This expression describes all the possible combinations of n things. For example, if two variables change in going from one state to the next, any of the four combinations may exist for a short time during the transition period, depending on which variable changes first. One of the four combinations is the previous state and another is the next state. The remaining two combinations may correspond to other states in the machine or don't care states. These two combinations can cause race outputs if they are included in an output function encirclement. In general, for n-variable changes, $2^n - 2$ states have to be checked for the possibility of producing critical race outputs; and any state in the machine may produce such outputs.

Output maps are the easiest description of the output function to use when checking for critical race outputs. For example, the ASM chart in Figure 4.69 describes a simple 3-state machine. A proposed reduced dependency state assignment is described by the state assignment map, and the output function for IH02 is described by an output map. Transitions from (a) to (b) and from (b) to (a) involve two variables; therefore, two states for each transition have to be checked for the possibility of race outputs. In the transition from state (a) to state (b), the possible intermediate state paths to be checked are indicated by arrows on the state assignment map shown in Figure 4.70.

As indicated, there may be an unwanted output caused by the intermediate transition from (a) to (c) to (b), because state (c) produces the IH02 output. The same arguments hold for the transition from (b) to (a) which may also cause a race output on IH02. Figure 4.71 shows the same state machine with a race-free state assignment. There are no states to check because all transitions are single-variable transitions.

Figure 4.72 compares the implementation of this example for both assignments. In this simple situation, the race-free assignment yields a marginally more complex implementation. Generally speaking, the

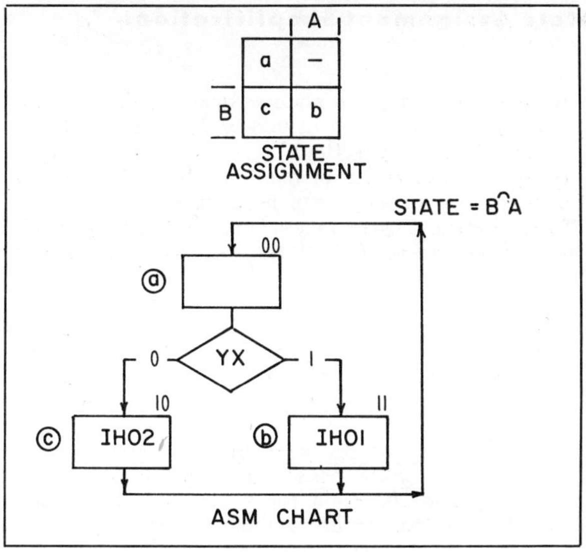

Figure 4.69 A State Assignment with an Output Race on IH02.

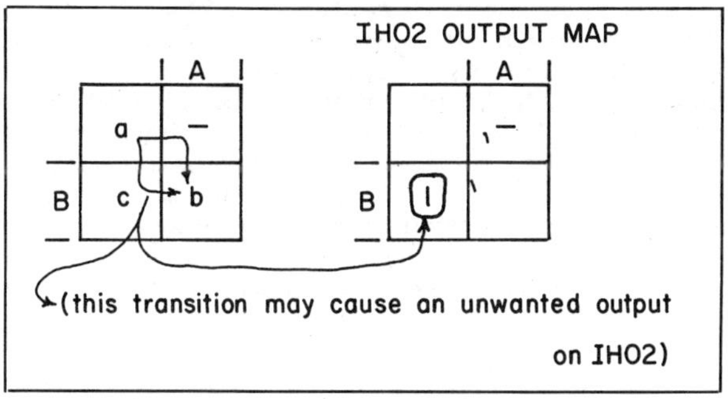

Figure 4.70 State Assignment Races.

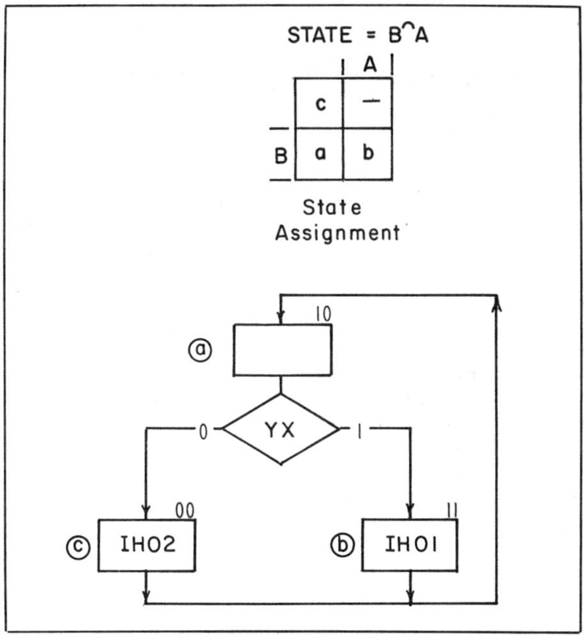

Figure 4.71 A Race-Free State Assignment.

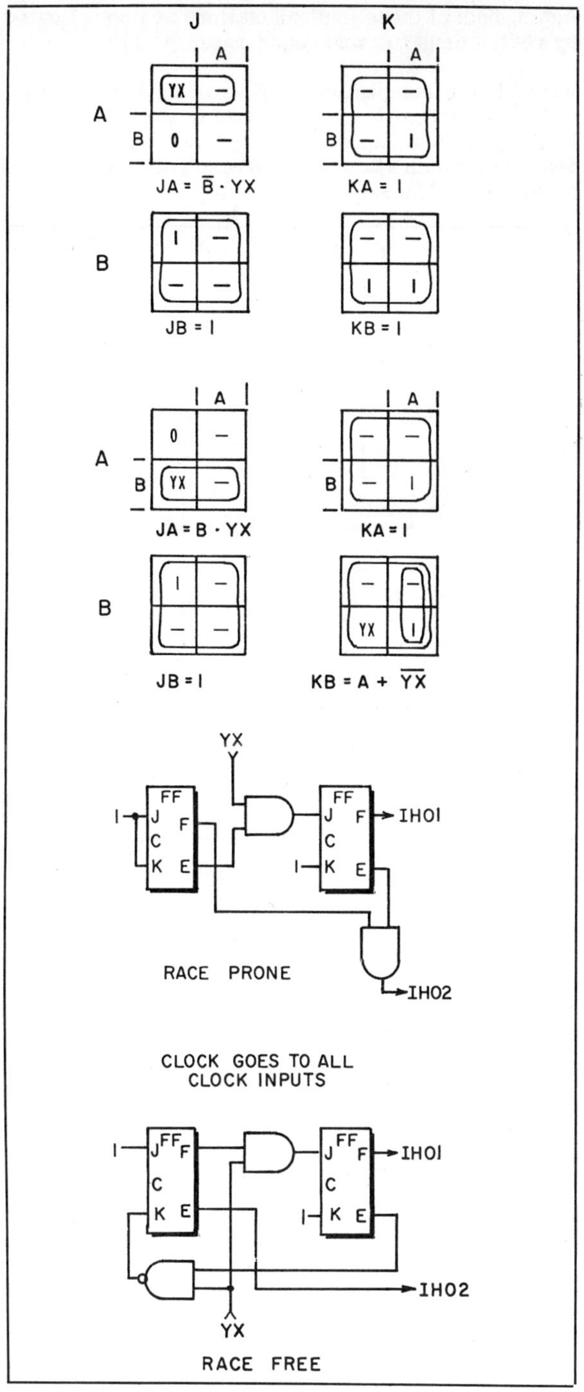

Figure 4.72 Implementation of Figure 4.69 and Figure 4.71.

race-free restrictions usually complicate the machine. The state assignment for race-free outputs sometimes involves adding extra state variables or extra states. Figure 4.73 describes a state machine which requires more than two variables to achieve a race-free assignment for three states. One possible assignment is given. Figure 4.74 describes the same state machine with one added state, which also has race-free outputs.

Again, both of these implementations are complicated by restrictions to avoid output races.

With all the complications of avoiding output races, it is usually best to avoid critical immediate outputs. In Chapter 6 on linked machines, some methods are described which allow immediate outputs to be isolated to small machines.

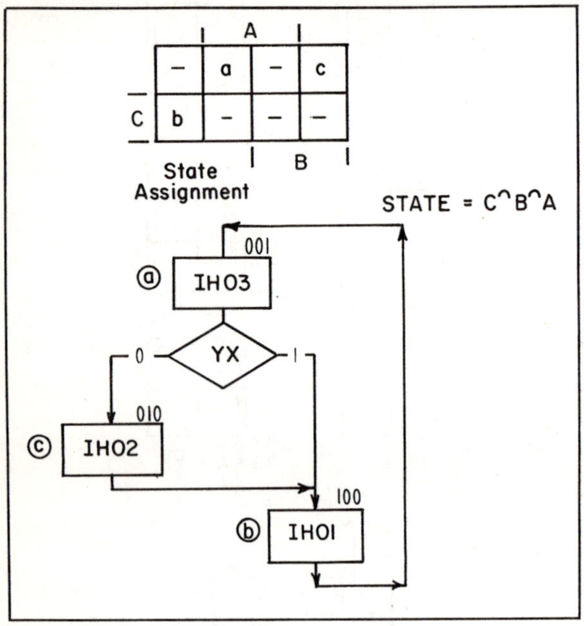

Figure 4.73 A Race-Free Assignment Using Extra State Variables.

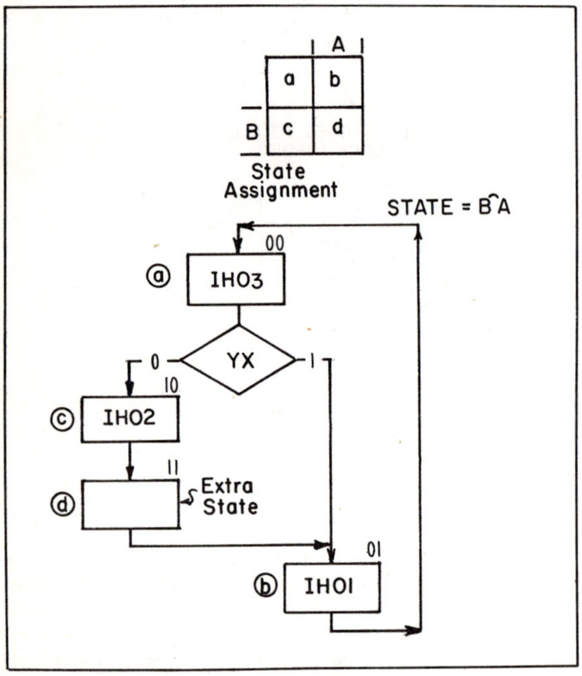

Figure 4.74 A Race-Free Assignment Using Extra State Variables and One Added State.

State Assignment Simplifications

The complete study of state assignment is a complex field beyond the scope of this book. This section has provided some practical guidelines without being overly mathematical. In most practical applications, however, the process can be reduced even further. The following suggestions can be verified by careful thought regarding the state assignment discussion.

States with asynchronous qualifiers must be assigned race free codes at the expense of all other simplifications. With due respect for rigorous methods, a race free assignment can be found by trial and error if the observation is made that two states reached by link paths separated by one asynchronous qualifier must have adjacent state codes. This condition is both necessary and sufficient for a race free assignment regardless of any other variables which may change. For example, consider an assignment for Figure 4.61 if YX is asynchronous. If YY=1, the transition of state (a) to state (c) does not depend on YX and is a non critical assignment. However, when YY=0, YX separates two link paths leading to state (a) and to state (b). These two states must have adjacent codes to be race free. No other restrictions are imposed on the assignment. With a little practice this simple rule is quite effective. An ASM chart must be constructed such that a race free assignment is possible. For example, any odd number of states linked in a loop by asynchronous qualifiers can not be made race free because they can not be made adjacent.

To simplify gating, direct state transitions should be given adjacent codes and conditional state transitions should be given reduced dependency assignments.

Conditional outputs should not be a function of asynchronous qualifiers because their time duration will be unknown.

Check for critical output races after a first try at an assignment and then converge on a final assignment by successive trials. It should only take a few tries. Better yet, avoid critical output functions by making all functions clocked.

A good state assignment is a key factor in good logic design. Since so much of the design work depends on the assignment, be sure that as much as possible is anticipated before proceeding with the gate implementation. In Chapter 5, some of the worries over gate minimization can be ignored; but, asynchronous qualifiers can still cause races.

4.9 HAZARDS

This section describes false outputs, called HAZARDS, that are caused by the gate configuration alone. A technique for avoiding hazards is described

Even though the states of an ASM chart are given codes to properly avoid races, there still remains a possibility that the gate implementation of immediate outputs may produce false outputs. Such false outputs, which are characteristic of the gate configuration rather than the logic function itself, are called HAZARDS. Figure 4.69 describes a very simple circuit which has a hazard caused by a delay through the inverter, gate 4. This hazard occurs even though only a single input changes and in spite of the fact there may be no delay at all in gates 1, 2 and 3. The map of the function in Figure 4.75 indicates that the hazard occurs during the transition between two input states which should both produce a 1 output on IHO. The generation of the hazard output is seen by observing the gate implementation. When Y1, Y2 and Y3 are equal to 1, the output of gate 1 is 0 and the output of gate 2 is 1, producing a steady 1 output for IHO. When Y2 goes to 0, there is some delay before the output of the inverter can change to 1. During this short time, the 0 output of gate 4 makes the output of gate 1 equal to 0, and the 0 input from Y2 makes the output of gate 2 equal to 0. Thus, both inputs to gate 3 are 0; and the output of gate 3 goes to zero until the output of gate 4 goes to a 1, allowing gate 1 to go to a 1 and returning the output of gate 3 to its proper stable state. The hazard output is shown on a time-line presentation below the map.

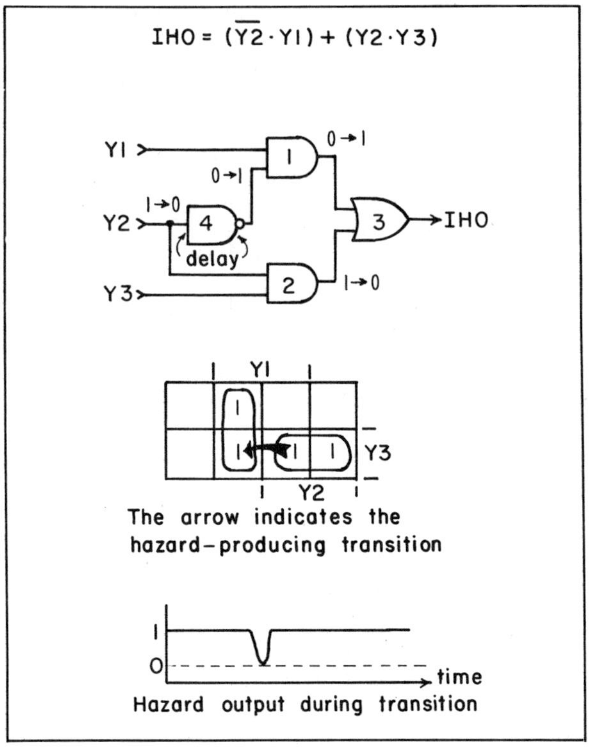

Figure 4.75 A Simple Output Circuit with a Hazard.

A hazard can occur during any of the four possible transitions of a single output. Figure 4.76 summarizes these hazards by giving a simple time-line example of each transition hazard.

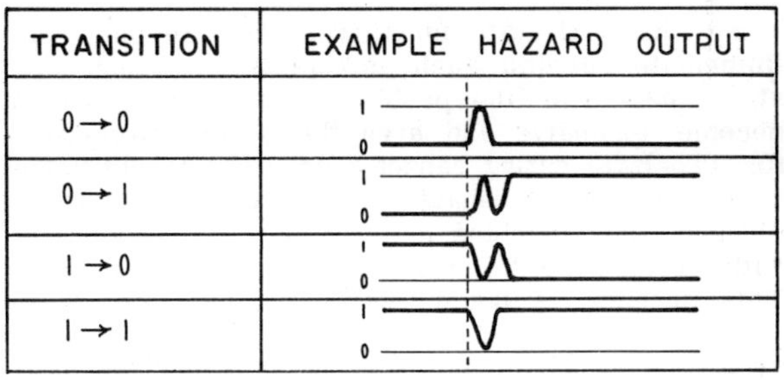

Figure 4.76 An Example of a Hazard.

Since the hazard outputs are much like the race outputs covered previously, in that they both occur during a transition, they can only cause problems in a critical immediate output. Hopefully, there are only a few such outputs. The hazards which may exist in these outputs may be far more obscure than in the previous example. It would be desirable to have a convenient means to detect the hazards and also to avoid them. So far only one general proof has been made which allows detection and correction of hazard outputs. However, the proof requires that the gate implementation be restricted to two-level gating ignoring input inverters. If such a restriction is a suitable alternative to careful checking of the possible gate delays, it may be used with complete success.

The theorem for a hazard-free design states that a two-level circuit implemented as a sum of products (the OR of AND gates) will be free of all transition hazards if it is free of hazards for the 1-to-1 transitions. This theorem simplifies the problem to one situation for which the hazards are recognizable and correctable.

A 1-to-1 transition occurs in an output map for any pair of a set of inputs which both produce a 1 output. Normally the pair which produce no race outputs have only one variable changing. As shown, even a single variable transition can produce a hazard where a delay through an inverter makes the complement of an input incorrect for a short period of time. Such situations occurring in two-level gating are easily recognized by 1 output states, which are a unit distance apart and encircled separately on a map of the output function. The change from one encircled area to another then indicates the change of gates selected by the true and the complement of an input.

The procedure for eliminating hazards in two-level gating consists of including all 1-to-1 transitions which are a unit distance apart in at least one encirclement. For example, in the previous illustration of a circuit with a hazard, the map had two encirclements with two 1's a unit distance apart. The hazard appeared during the transition between these two 1's, just as would be predicted. Figure 4.77 illustrates the elimination of the hazard by including these two 1's in a third encirclement. The added gate 5 serves to define the output during the transition. By simple extensions of this procedure, the hazards may be eliminated from any two-level gate implementation.

When the next-state function is constructed from immediate outputs, such as was done in unclocked state machines, the problems caused by hazards become extensive and even the simple procedure for two-level gating cannot adequately handle some of the erroneous behavior which results. Each unclocked state machine must be handled individually with great care to avoid delay-generated errors. This problem is not discussed because it tends to be specialized and because clocked state machines avoid the problem. In addition, it is not essential to the study of algorithmic state machines.

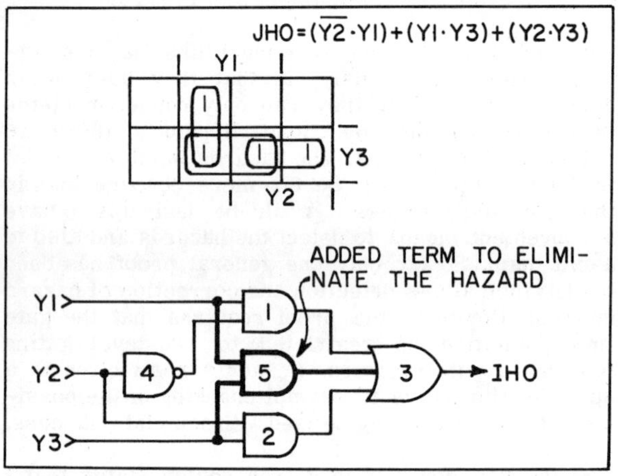

Figure 4.77 The Application of Additional Gating to Eliminate the Hazard in a Simple Output Circuit.

4.10 SAMPLE IMPLEMENTATION

The Black Jack machine described in the end of Chapter 3 is implemented in this section as a class 4 conditional output machine. The design is discussed from the level of the ASM chart description up to the level of the gate implementation. There are three areas in the design process:

1. State assignment
2. Next-state function
3. Output Function

State Assignment

Figure 4.80 is an ASM chart description of the machine to be designed. The state assignment which is shown in this chart is determined by applying the steps in the procedure described in Section 4.8. Accordingly, each state with asynchronous qualifiers is checked for the existence of a race-free assignment. Thus state (a) must be a unit distance from state (c), and state (g) and state (h) must be a unit distance from state (b). The consideration of state (d) is assisted by constructing a next-state dependency map, because the exsistence of a race-free assignment is not as evident as in the last three states. Figure 4.72 shows the dependency map for state (d), assuming separate change vectors for each next state. Only the change vector b listed in Figure 4.72 is a function of *YCRD, meaning that a race-free assignment for state (d) is possible. However, the change vectors are formed on the basis of four variables and only three are really required to code the six states in the ASM chart. A three-variable race-free assignment for state (b) would still have only one change vector dependent on the asynchronous qualifier. Since YG16 separates change vector b from the three other change vectors, as long as one variable changes for vector b any combinations of variables can change for the remaining three change vectors; and a race-free state assignment having only three variables should be possible. At this point, a race-free assignment using three state variables is attempted.

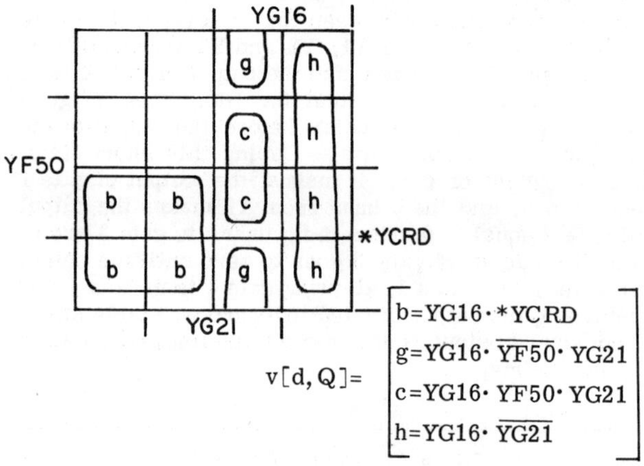

Figure 4.78 The Next-State Dependency Assuming Independent Change Vectors.

The ASM chart states are given state assignments to follow the restrictions imposed by races. The first state code is given to state (d). For convenience, let (d) = 101. A single variable change to state (b) is required. Let (b) = 111. States (g) and (h) must also be a single-variable change from state (b). Let (g) = 011 and (h) = 110. Since a reduced dependency assignment is not likely for state (c), let state (c) = 001 to reduce the state locus from (d) to (c) and (c) to (d). This choice forces state (a) to equal 000 in order for it to be a unit distance from state (c). The last transition, from (b) to (a), involves all three variables, but this seems to be a penalty for having restrictions elsewhere. The proposed state assignment is summarized by the map in Figure 4.73.

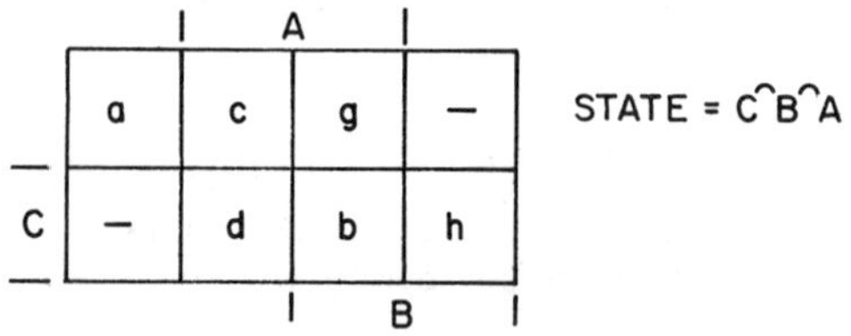

Figure 4.79 The Proposed State Assignment.

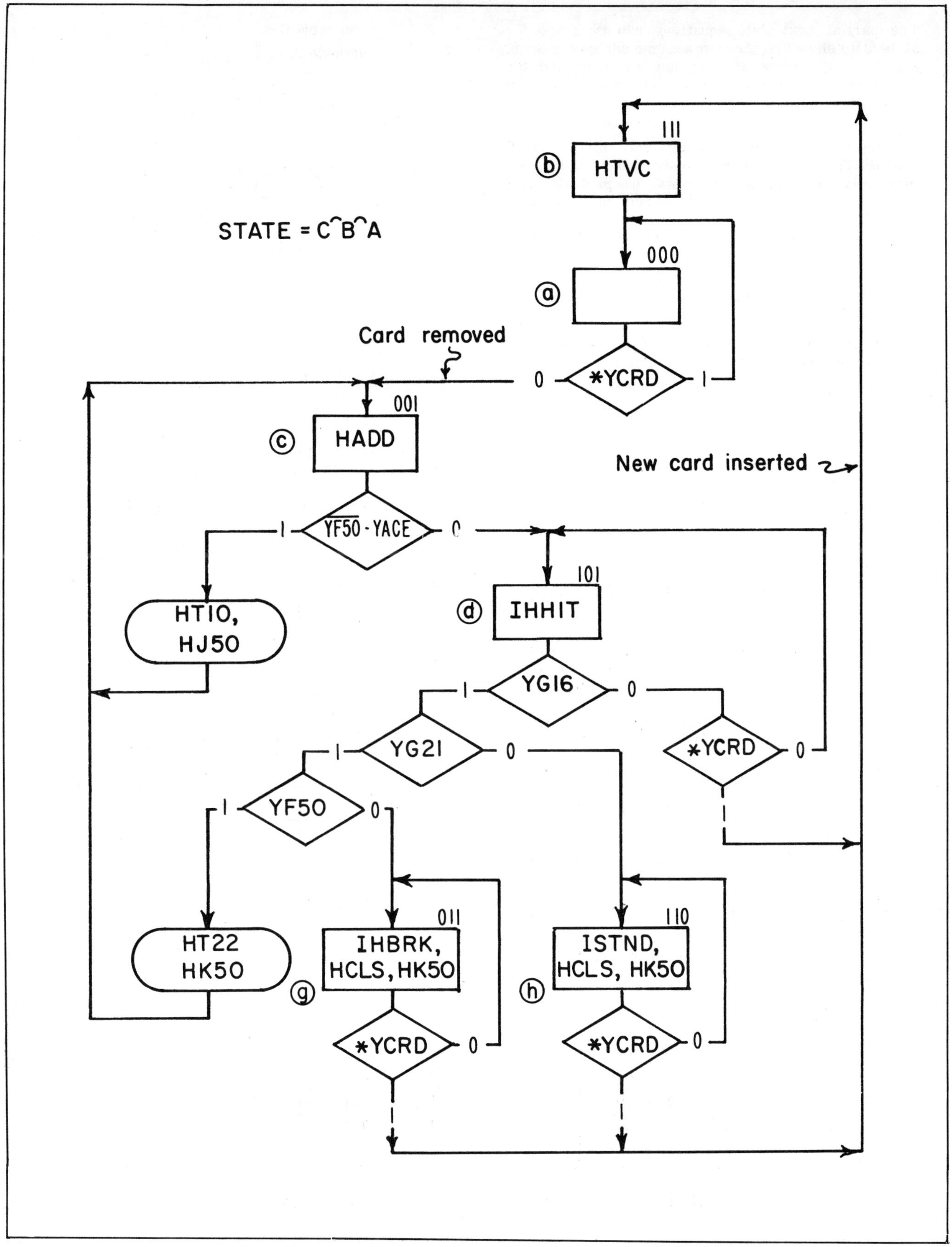

Figure 4.80 The Black Jack Control Algorithm Described as a Conditional Output, Conditional Transition (Class 4) Machine.

The partial next-state equations can be found for state ⓓ to show that the state assignment is race-free. Figure 4.81 shows the dependency map and the transition vector encirclements to generate the partial next-state equations shown.

For the purpose of this example, only IHHIT, IHBRK and IHSTD are immediate instructions; and, since they only drive indicator lights, no problems are created by possible erroneous outputs during transitions. If there were critical immediate outputs, races might be caused by the three-bit transition from ⓑ to ⓐ.

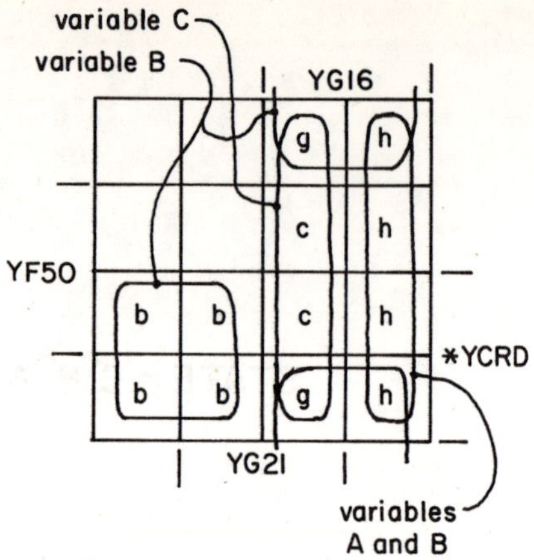

$$g_A\,[d,\,Q] = YG16 \cdot \overline{YG21}$$

$$g_B\,[d,\,Q] = (\overline{YG16} \cdot {*}YCRD) + YG16\,(\overline{YF50} + \overline{YG21})$$

$$g_C\,[d,\,Q] = YG16 \cdot YG21$$

No race since only variable B depends on *YCRD.

Figure 4.81 Checking State ⓓ for Races.

Next-State Function

The input functions to generate the next states using JK flip-flops are constructed as follows:

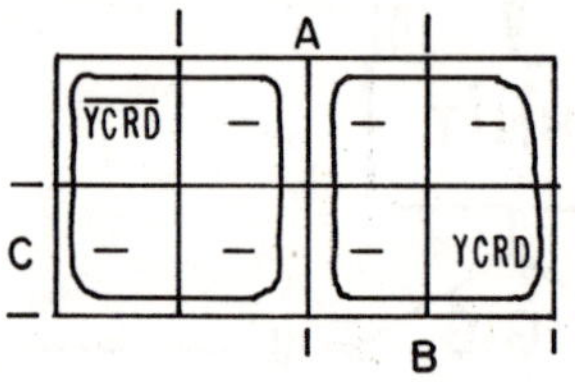

$JA = (\overline{YCRD} \cdot \overline{B}) + (YCRD \cdot B)$

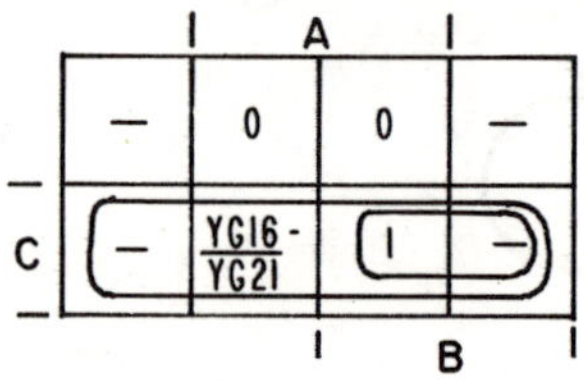

$KA = (C \cdot B) + (YG16 \cdot \overline{YG21} \cdot C)$

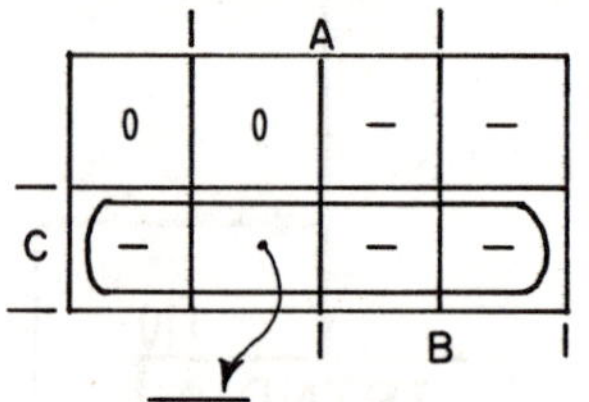

$JB = C \cdot ((\overline{YG16} \cdot YCRD) + YG16(\overline{YF50} + \overline{YG21}))$

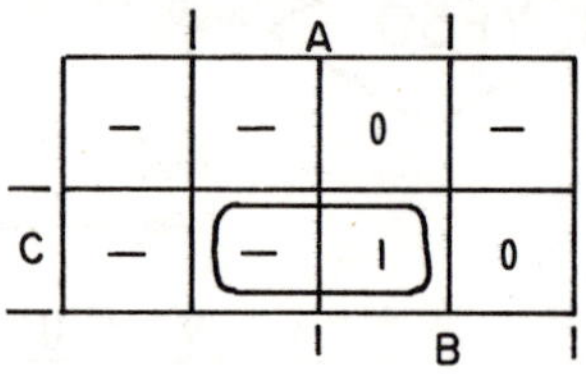

$KB = C \cdot A$

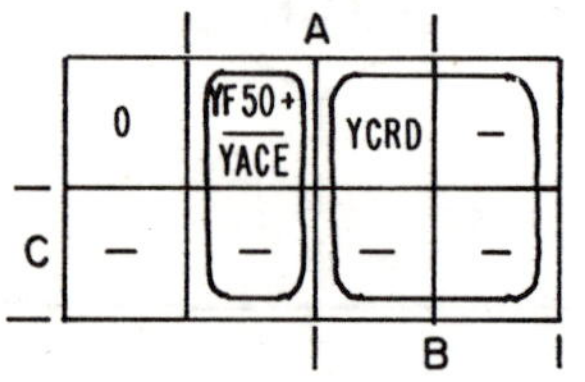

$JC = ((YF50 + \overline{YACE}) \cdot A \cdot \overline{B}) + (B \cdot YCRD)$

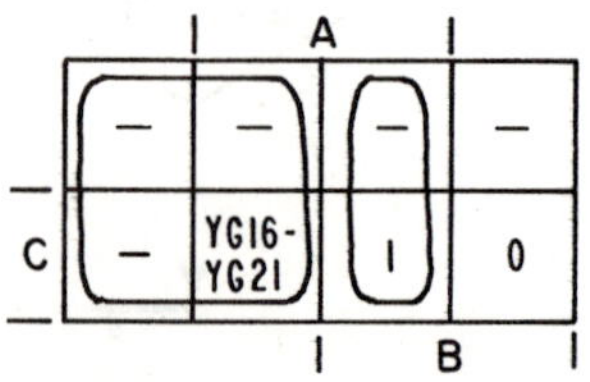

$KC = (\overline{B} \cdot YG16 \cdot YG21) + (A \cdot B)$

Output Function

The output functions are determined by the following maps. Notice that in forming output equations some outputs are natural extensions of other outputs as indicated by those used in the right hand side of the equations. This observation saves redundant gating.

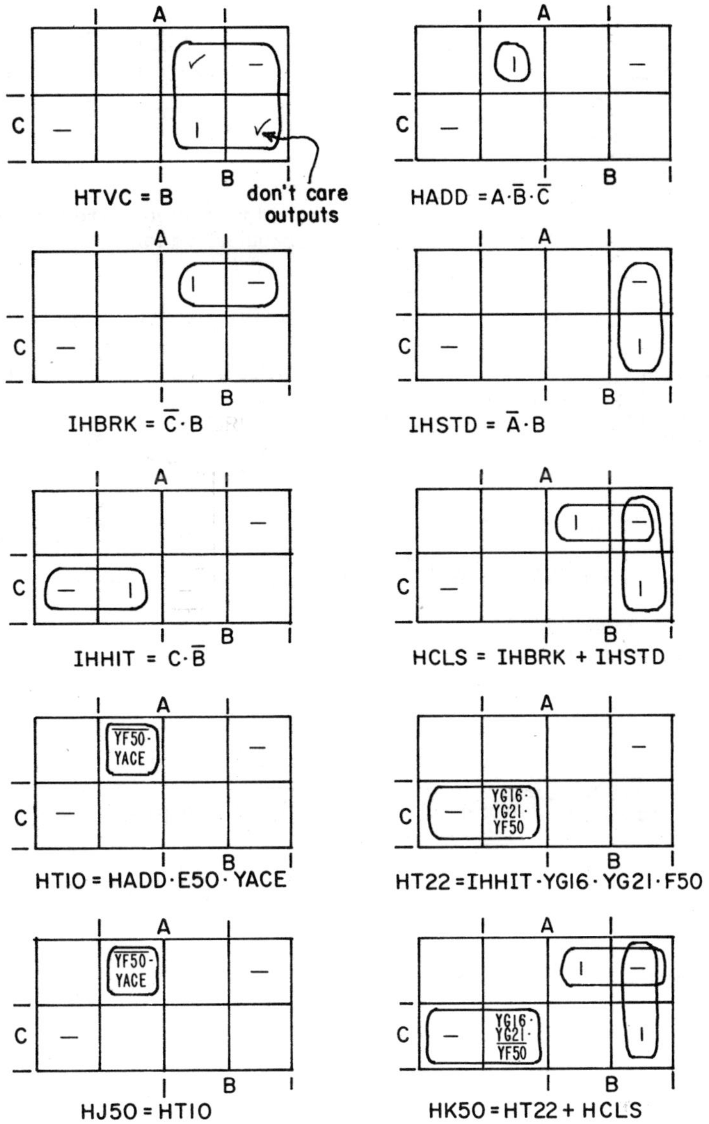

Gate Realizations

The equations for the state variable inputs and the outputs have to be converted into gate realizations. For this problem, assume that the following gates are available:

2-, 3- and 4-input NAND gates,

2-input NOR gates, and

2-input AND gates.

An inverter is made from a 2-input NOR or NAND gate.

The next-state variable inputs are first converted to gate implementations as follows:

$JA = (\overline{YCRD}\cdot\overline{B}) + (YCRD\cdot B) = \overline{\overline{(\overline{YCRD}\cdot\overline{B})}\cdot(\overline{YCRD\cdot B})}$

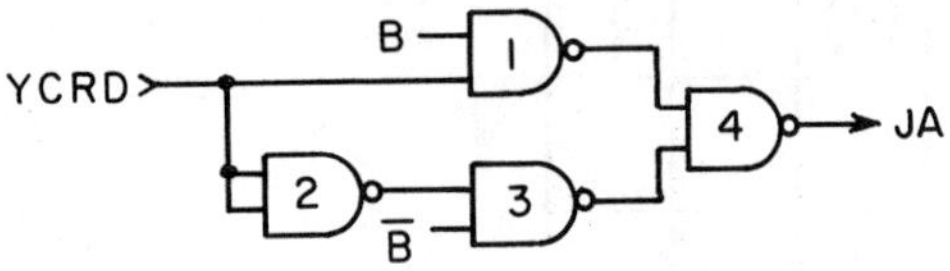

$KC = (\overline{B}\cdot YG16\cdot YG21) + (A\cdot B) = \overline{\overline{(\overline{B}\cdot YG16\cdot YG21)}\cdot(\overline{A\cdot B})}$

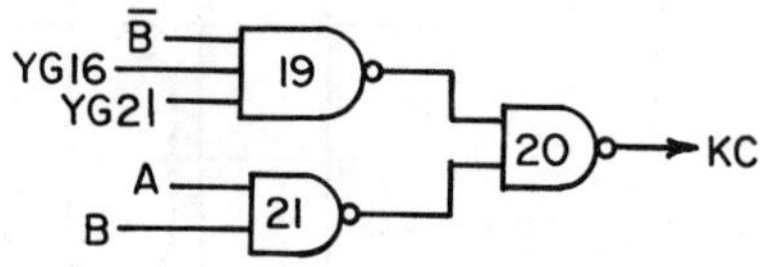

$KA = (C\cdot B) + (YG16\cdot YG21\cdot C) = \overline{\overline{C\cdot B}\cdot\overline{YG16\cdot YG21\cdot C}}$

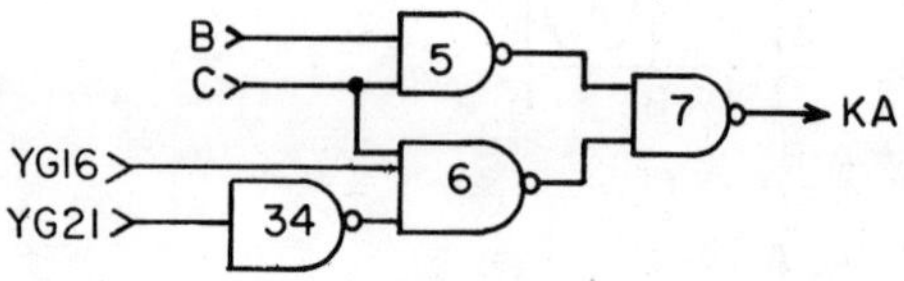

The output functions are converted to gates in a similar manner, as follows:

$HADD = A\cdot\overline{B}\cdot\overline{C} = \overline{\overline{A}+\overline{\overline{B}\cdot\overline{C}}}$

$HT10 = HADD\cdot\overline{YF50}\cdot YACE = \overline{YF50+\overline{HADD\cdot YACE}}$

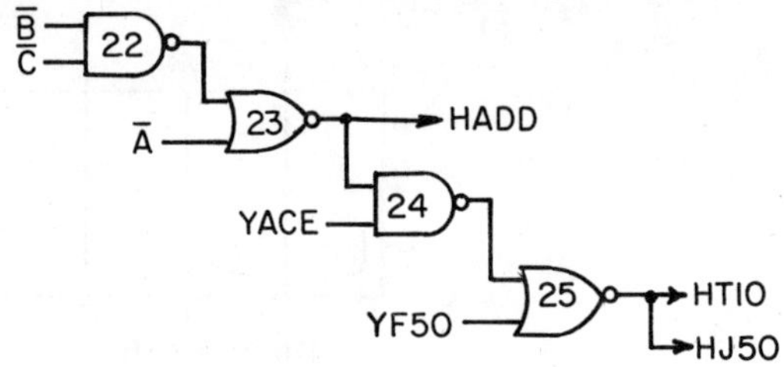

$JB = C\cdot((\overline{YG16}\cdot YCRD) + YG16\cdot(\overline{YF50}+\overline{YG21}))$

$= \overline{\overline{C} + \overline{(\overline{YG16}\cdot YCRD) + \overline{\overline{\overline{YG16}}+(YF50\cdot YG21)}}}$

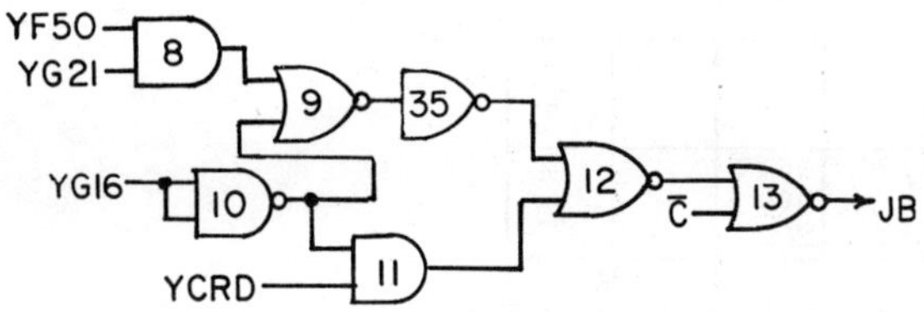

$IHBRK = \overline{C}\cdot B$

$IHSTD = \overline{A}\cdot B$

$HCLS = IHBRK + IHSTD = \overline{\overline{IHBRK + IHSTD}}$

KB= C·A

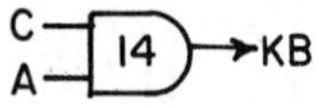

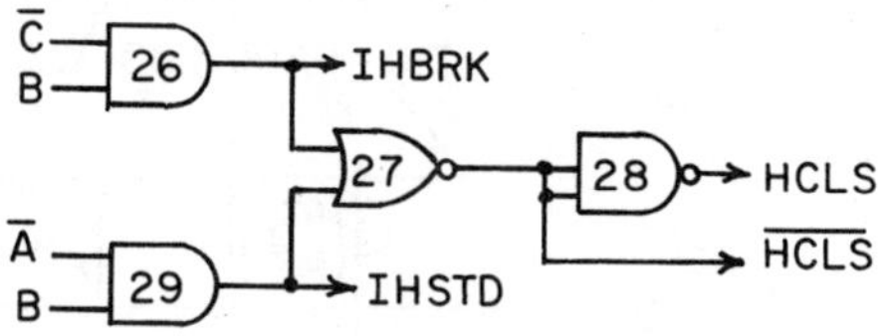

$JC = ((YF50+\overline{YACE})\cdot(A\cdot\overline{B})) + (B\cdot YCRD))$

$= \overline{\overline{\overline{\overline{YF50}\cdot YACE}\cdot A\cdot\overline{B}}\cdot\overline{B\cdot YCRD}}$

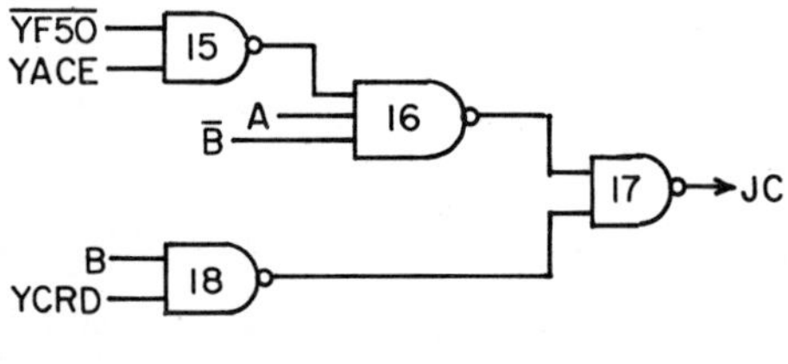

$IHHIT = C\cdot\overline{B}$

$HT22 = \overline{\overline{IHHIT\cdot YG16\cdot YG21\cdot YF50}}$

$HK50 = HT22 + HCLS = \overline{\overline{HT22}\cdot\overline{HCLS}}$

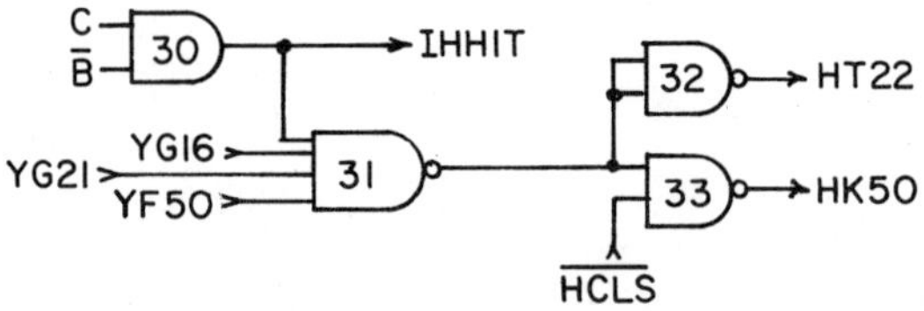

Figure 4.82 is a composite logic circuit for the Black Jack control.

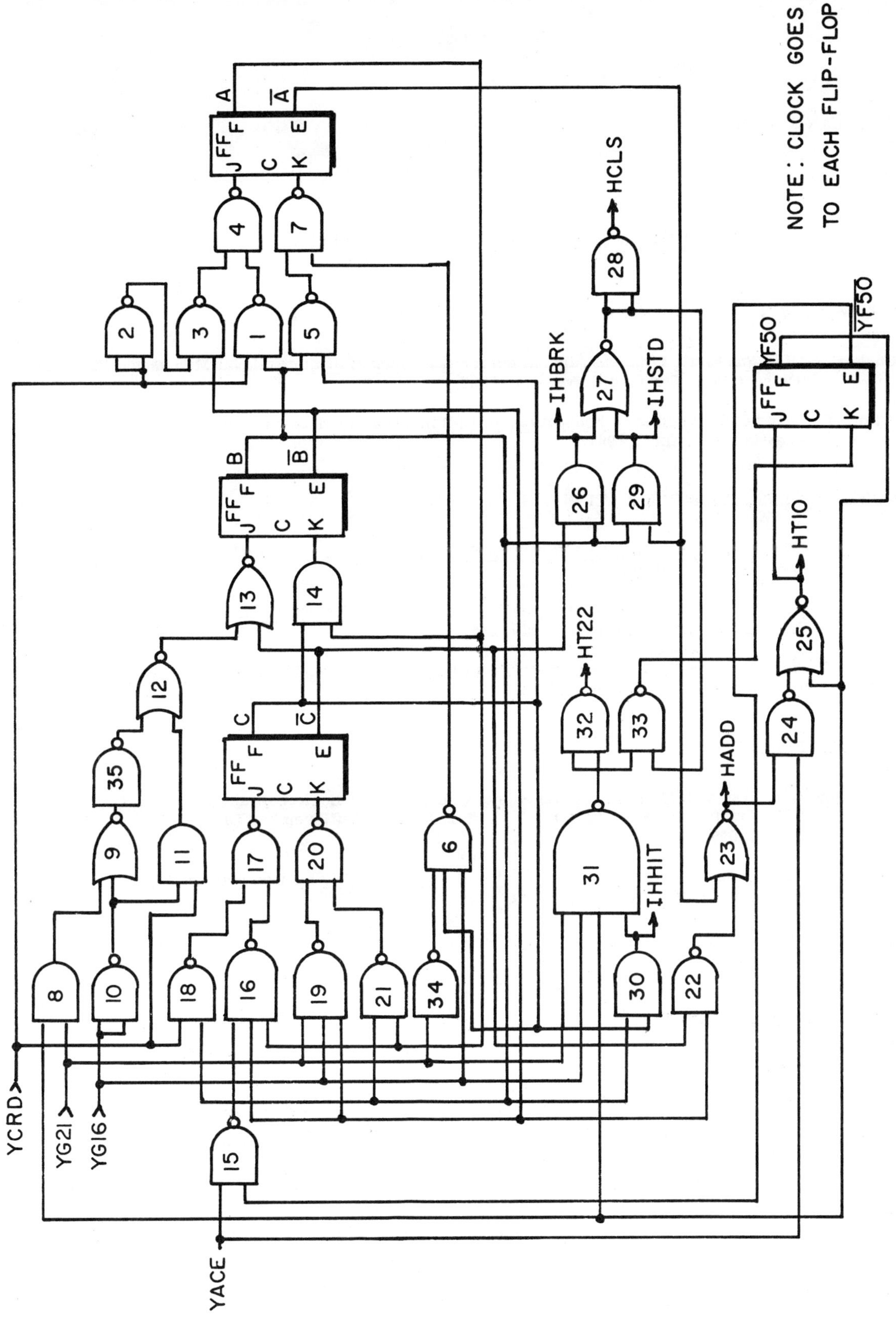

Figure 4.82 Complete Logic Circuit for the Black Jack control Module Including the 11-Point Flag.

REFERENCES

CHAPTER IV

BOOLEAN ALGEBRA

Boole, George: An Investigation of the Laws of Thought (New York: Dover Publications, Inc., 1951 1st ed., 1854).

McCluskey, E. J.: Introduction to the Theory of Switching Circuits (New York: McGraw-Hill Book Company, 1965), Chapter 3.

OPTIMIZATION

Gucker, G. R.: Stochastic Gradient Algorithms for Searching Multidimensional Multimodal Surfaces (Dissertation: Stanford University, 1969).

Wilde, D. J. and C. S. Beightler: Foundations of Optimization (Englewood Cliffs, New Jersey: Prentice-Hall, Inc., 1967).

QUINE-McCLUSKEY SIMPLIFICATION

McCluskey, E. J.: Introduction to the Theory of Switching Circuits (New York: McGraw-Hill Book Company, 1965).

McCluskey, E. J.: "Minimization of Boolean Functions," Bell System Tech. J. Vol. 35 (November 1956), pp. 1417-1444.

Quine, W. V.: "The Problem of Simplifying Truth Functions," Am. Math Monthly, Vol. 59 (October 1952), pp. 521-531.

MAP SIMPLIFICATION

Karnaugh, G.: "The Map Method for Synthesis of Combinatorial Logic Circuits," AIEE Trans. Commun. Electron., Part I, Vol. 72 (November 1953), pp. 593-599.

Maley, Gerald and John Earle: The Logic Design of Transistor Digital Computers (Englewood Cliffs, New Jersey: Prentice-Hall, Inc., 1963).

Wickes, William E.: Logic Design with Integrated Circuits (New York: John Wiley & Sons, Inc., 1968).

TOPOLOGICAL SIMPLIFICATIONS

Miller, Raymond E.: Switching Theory, Vol. 1: Combinatorial Circuits (New York: John Wiley & Sons, Inc., 1965).

Roth, J. P.: "Algebraic Topological Methods for the Synthesis of Switching Systems," Trans. Am. Math. Soc., Vol. 88 (July 1958), pp. 301-326.

MISCELLANEOUS SIMPLIFICATIONS

Slagle, James R.: "A New Algorithm for Generating Prime Implicants," IEEE Trans. Electron. Comput., Vol. C-19 (April 1970), pp. 304-310.

Tison, P., "Generalization of Consensus Theory and Application to Minimization of Boolean Functions, "IEEE Trans. Electron. Comput., Vol. EC-16 (August 1967), pp. 446-456.

SEQUENTIAL MACHINES

Dietmeyer: Logic Design of Digital Systems (Boston: Allyn and Bacon, 1971).

Miller, Raymond E.: Switching Theory, Vol. 2: Sequential Circuits and Machines (New York: John Wiley & Sons, Inc., 1965).

Phister, Montgomery, Jr. Logical Design of Digital Computers (New York: John Wiley & Sons, Inc., 1966), Chapters 5, 6.

CHAPTER V

SYNTHESIS FOR ROM-CENTERED DESIGN

5.0 THE READ-ONLY MEMORY

The advent of large scale integrated arrays has brought to the forefront an old and basic idea of using complete arrays to execute combinatorial logic functions. Heretofore these arrays were considered highly inefficient solutions even though the design was simple, primarily because of the great waste of circuits. Now as the cost per function and the size decrease, the ease of design has again become appealing. The arrays, most often called READ-ONLY MEMORIES or ROMs for short, have almost become a panacea for logic design problems. However, the reader should not be misled. The choice of using ROMs must be made carefully because it may not always be the best solution. This chapter will attempt to cover the basic ROM considerations building upon the results in the last chapter.

5.1 A ROM STRUCTURE AS COMBINATORIAL LOGIC

The basic ROM is a combinatorial circuit producing a number of outputs for each set of inputs. Logic circuits which were implemented with individual gates can be PROGRAMMED in a ROM by a process comparable to listing the desired outputs. The new concepts introduced in this section are the ADDRESS, the WORD, the OR-TIE array and ADDRESSING EFFICIENCY.

The Address and the Word

The read-only memory consists of two sections, the address decoder part and the memory part as shown in Figure 5.1. The address part provides a means of selecting a portion of the memory part called a WORD. The input to select a given word is called the ADDRESS of the word. An address usually consists of an n-bit code which can select a maximum of 2^n words. Words consists of fixed-length bit strings from the memory part and are characteristic of the particular ROM. The information (bit pattern) stored in the ROM is fixed in respect to the logic environment being placed there during manufacture or by a process external to the normal operation of the logic. Thus, the memory is READ ONLY with respect to the logic around it.

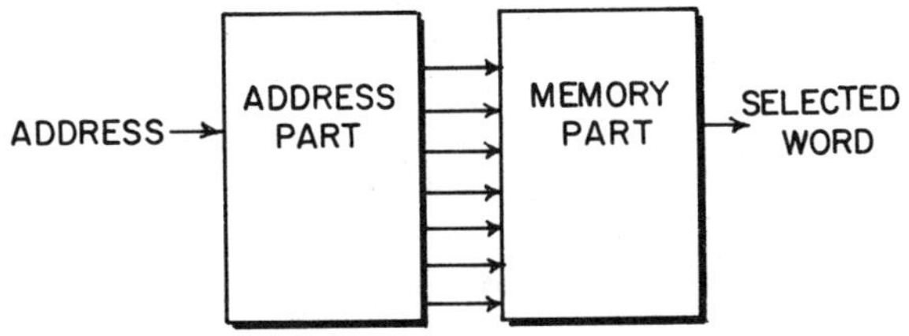

Figure 5.1 The Two Parts of a ROM.

To demonstrate how a ROM might be implemented, a pair of simple three-variable Boolean functions are generated by, first, forming eight address lines from the three input variables and then ORing the desired outputs into 2 bit output word. Two functions and the resulting circuit are shown in Figure 5.2.

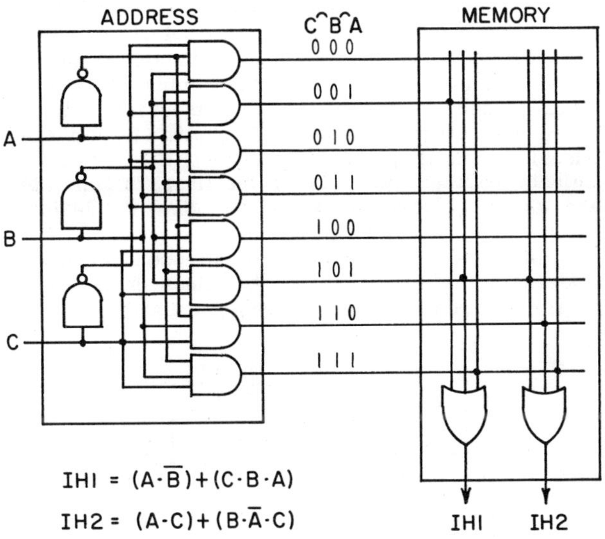

Figure 5.2 An Example Implementation of Two Simple Functions in a ROM Structure.

The OR-TIE Array

Rather than drawing all the inputs to the OR gates in the memory section, a simplified representation is adopted as shown in Figure 5.3. In this representation the OR inputs are indicated as one line

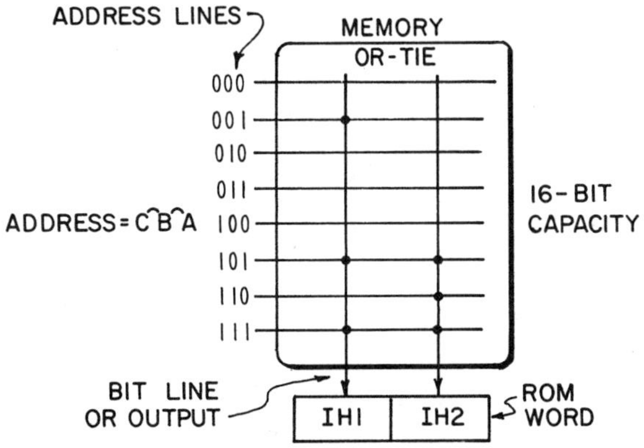

Figure 5.3 A Simplified Representation of a ROM.

called the BIT LINE, which is tied to the address lines by dots to indicate an input to the OR gates. The OR function is indicated by the internal reserved word OR-TIE. Each intersection of the bit and address lines is a possible location for a bit and thus the bit capacity of a ROM is the total number of intersections whether or not they are used. The OR output is indicated by arrows on the output lines.

Addressing Efficiency

The number of address lines required for a given number of bits stored depends upon the organization of the addressing. In the previous example the 8 possible bits for each bit output were selected by eight separate address lines. This organization is called one-dimensional. In Figure 5.4 the number of address lines is decreased to 4 and the number of bit lines increased to 4 in the process of forming a two dimensional organization. In one dimension, the C variable selects between two sets of bit lines to generate the desired output word. In the other dimension, variables A and B select one of four groups of bits for the bit lines. For each bit of an output word the addressing is four lines by two lines or six lines total compared to eight lines used previously.

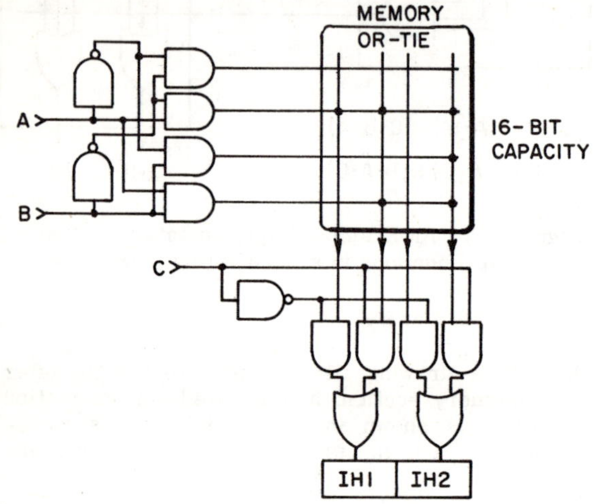

Figure 5.4 Two-Dimensional Addressing.

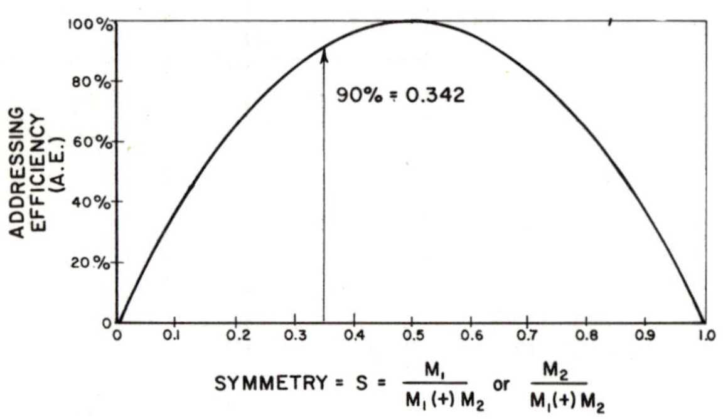

Figure 5.5 The Efficiency of a Two Dimensional Memory Word Addressing.

The relationship of the address lines to the bits stored may be expressed in terms of the percentage of the maximum number of bits addressable with a given number of address lines in two dimensions. If M_1 and M_2 represent the number of address lines in two dimensions, then M_1 (·) M_2 represents the total number of bits addressed. The maximum number of bits addressed for a given sum of M_1 (+) M_2 is $\left[\frac{M_1 (+) M_2}{2}\right]^2$ or the number of bits when both address dimensions are equal.[1]

The number of bits addressed expressed as a percentage of the maximum number of bits addressable with a given number of address lines is called the ADDRESS EFFICIENCY (A.E.).

$$\text{A.E.} = \frac{M_1 (\cdot) M_2}{\left[\frac{M_1 (+) M_2}{2}\right]^2} (\cdot)\ 100\% \qquad (1)$$

In terms of the normalized quantity S, called the SYMMETRY of the addressing, the equation is

$$\text{A.E.} = 400S(1(-)S) \text{ where } S = \frac{M_1}{M_1 (+) M_2} \qquad (2)$$

This equation is plotted in Figure 5.5. From this graph a 10 by 10 selection is 100% efficient, while a 5 by 20 selection is only 64% efficient.

In summary of combinatorial logic in ROMs, the information stored is analogous to a tabular description and symmetry in the address dimensions sets the efficiency in selecting this information. A ROM is most useful for very complex functions rather than for simple ones. Code conversions and character generation are good examples of complex transform-type functions suitable for ROM implementation. A ROM is described by specifying the exact bit pattern contained in the memory section and the address required for the selection of a specific word. In the next section, the structure of a ROM-centered algorithmic state machine is described in terms of the information storage in the ROM.

5.2 SEQUENTIAL LOGIC IN ROMS

A combinatorial ROM is used to implement class 2, 3, and 4 state machines by storing the next-state function and the output function in the ROM array and providing some small portion of external logic to execute the stored information. The storage is characterized by the WORD STRUCTURE. Three ROM machine word structures are described in this section and are called LINK-PATH ADDRESSABLE, STATE-QUALI-

---

[1] $C = M_1 (\cdot) M_2$; $L = M_1 (+) M_2$; $\frac{dC}{dM_2} = L - 2M_2 = 0$

$M_2 = \frac{L}{2} = \frac{M_1 (+) M_2}{2}$; $M_1 = M_2$

$\therefore C_{max} = \left(\frac{L}{2}\right) (\cdot) \left(\frac{L}{2}\right) = \left[\frac{M_1 (+) M_2}{2}\right]^2$

FIER PAIR ADDRESSABLE and VARIABLE FORMAT ADDRESSABLE. New concepts introduced in this section are LINK, TEST, and INSTRUCTION word parts, AND-TIE array, DIRECT and ALTERNATE ADDRESSES, ASSUMED ADDRESSES, X-LINKS, LINK STRINGS, FIXED and VARIABLE FORMATS.

5.2.1 Link-Path Addressable

The link-path addressable word structure is based on storing the next state and the output for each link path in the ASM charts. The next-state portion of the ROM word is called the LINK PART, while the output portion of the ROM word is called the INSTRUCTION PART. Each address is a function of the present state and the qualifier inputs and is called a LINK-PATH address. In general any machine described by an ASM chart can be implemented with this structure, called a LINK-PATH ADDRESSABLE ROM. Figure 5.6 describes a link-path addressable ROM and illustrates a possible link path on an ASM chart.

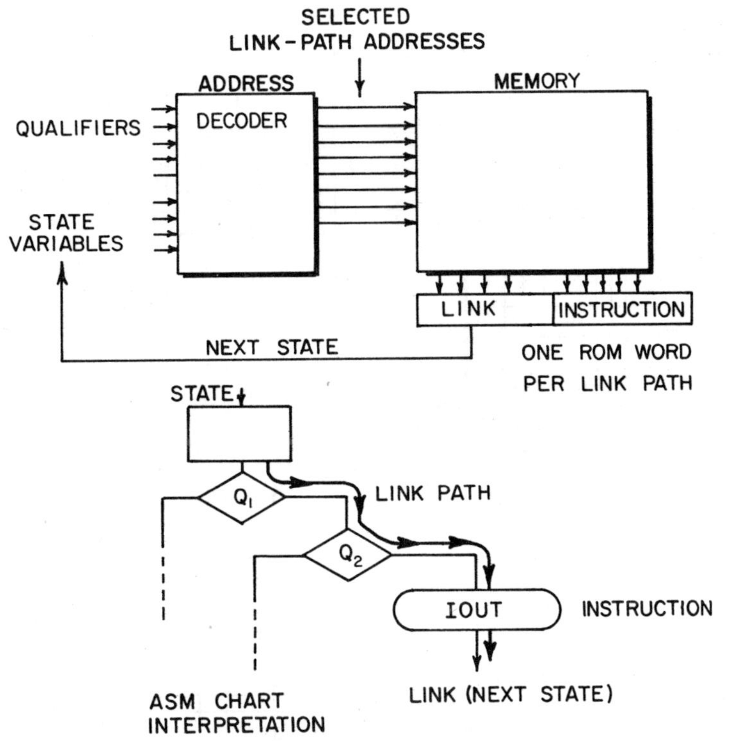

Figure 5.6 A Representation of a Link-Path Addressable ROM.

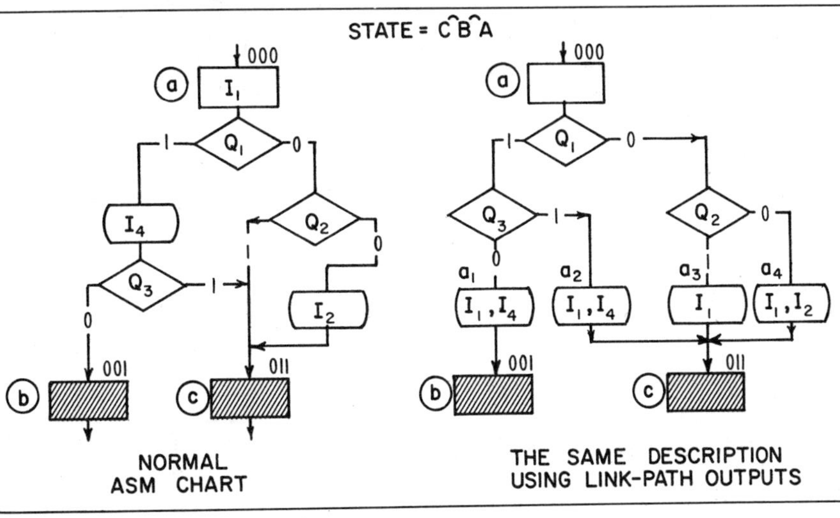

Figure 5.7 A Comparison of Normal and Link-Path Output Descriptions in an ASM Chart.

The outputs in a link-path addressable ROM are link-path outputs rather than state outputs. The outputs correspond to conditional outputs on the exit paths from each ASM chart block. Any output in an ASM block can be expressed as an exit output as illustrated by Figure 5.7. The normal description is given on the left. The equivalent link-path outputs shown on the right are obtained by moving all the outputs through all the link paths toward the exits and taking the sum of the outputs resulting for each exit as the new link-path outputs. Each link-path exit will be indicated by an additional subscript on the state name as shown.

This state can be implemented with the ROM as shown in Figure 5.8, which shows only a portion of the link-path address lines for the entire machine. The entire machine consists of the words for all the states.

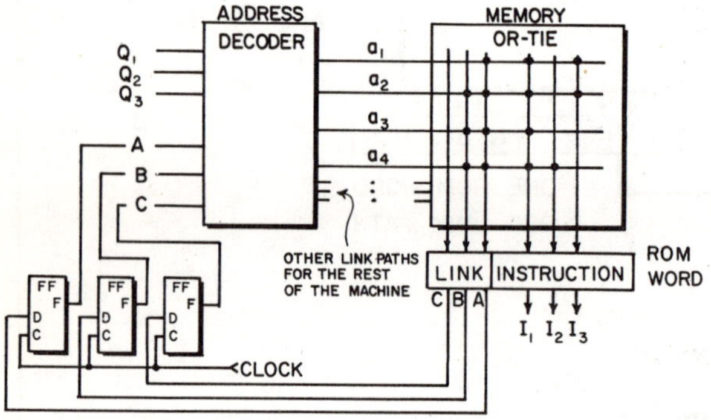

5.8 An Implementation of One State with a Link-Path Addressable ROM.

The address decoder can be implemented with ordinary combinatorial reduction and gates. For example the portion of the link-path decoder described in Figure 5.7 might be implemented as shown in Figure 5.9. However, this implementation provides no general solution to the decoder. Figure 5.10 gives another implementation of the decoder using an array for the AND gate inputs. The state and inputs for each link-path address are determined by simply connecting the appropriate AND inputs. As with the memory portion this decoder may be represented as shown by an array called AND-TIE, which has single lines for the collective AND inputs, and by arrows to indicate the AND gate outputs as shown in Figure 5.11. The link-path addressable ROM may then be considered as two arrays: an AND-TIE array for the addresses and an OR-TIE array for the memory. These two arrays are shown together with the state flip-flop in Figure 5.12.

The problem with a link-path addressable ROM is that the information to describe the operation of an algorithmic state machine is stored in two different types of arrays, and if there are very many qualifiers, the size of the AND-TIE array can easily be as large as the OR-TIE array. In addition, only a few of the possible combinations for link paths are used. It is desirable to find a word structure which allows description of the ASM chart in one type of array. Such structures are described next.

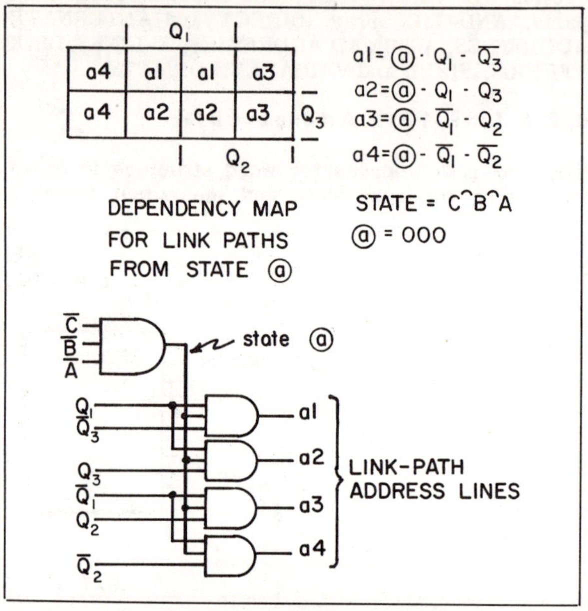

Figure 5.9 A Portion of the Address Decoder Implemented in Gates.

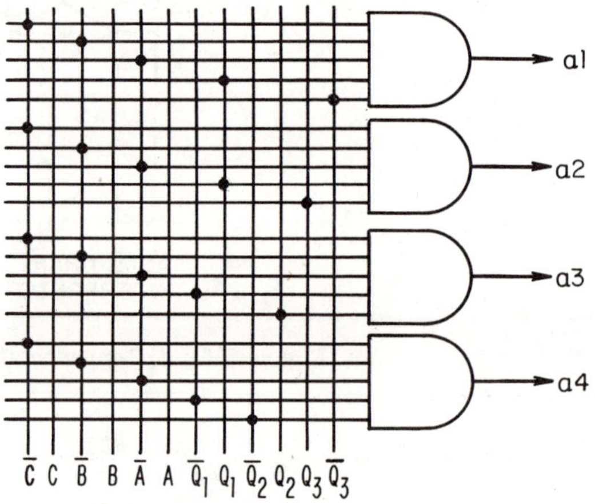

Figure 5.10 A General Decoder Implementation.

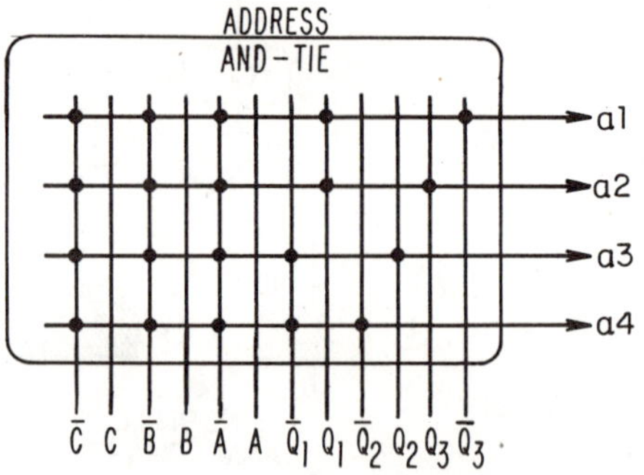

Figure 5.11 The AND-TIE Simplified Representation for an Address Decoder.

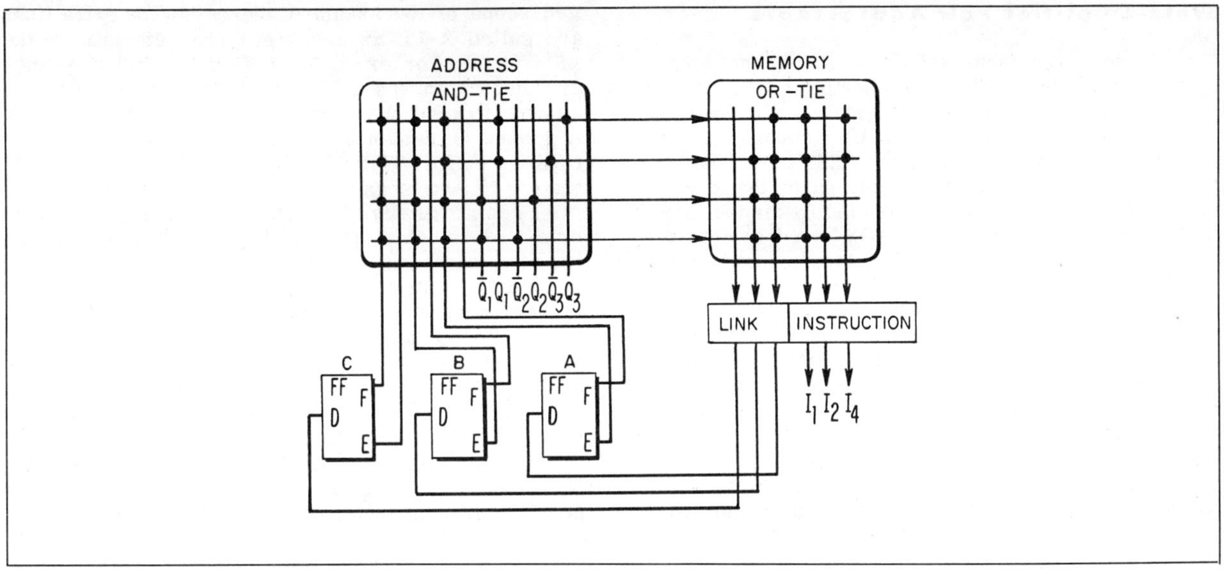

5.12 The Link-Path Addressable ROM as Two Interconnected Arrays.

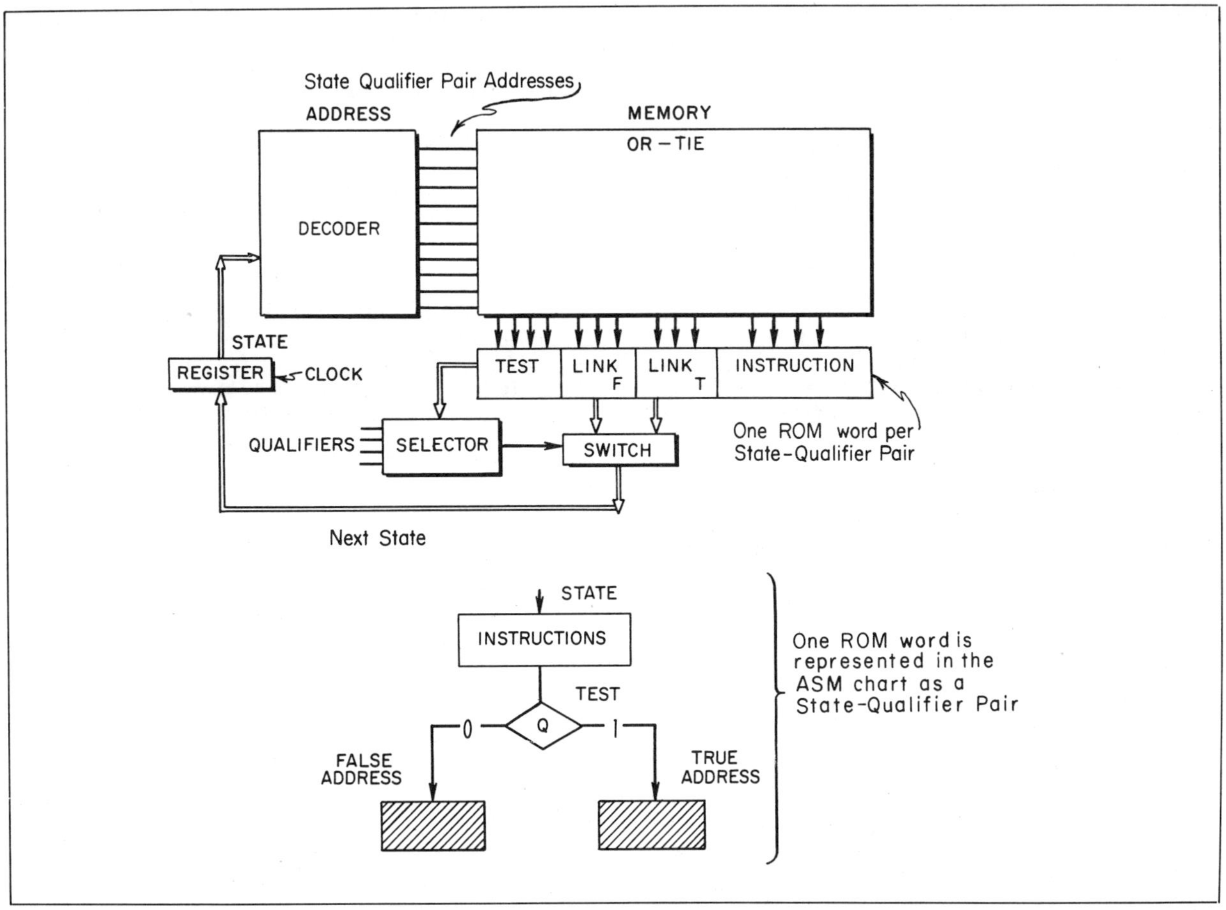

Figure 5.13 State-Qualifier Pair ROM Address Structure.

State-Qualifier Pair Addressable

One way to store information in the OR-TIE array only is to restrict the ASM chart description to a class 3 machine with one qualifier per state. A new portion of the ROM word names the qualifier chosen for each state. This portion is called the TEST part. The link part now has two next states chosen on the basis of the outcome of the qualifier selected by the test part. The instruction part of the ROM word selects the desired state outputs as was done before for conditional outputs. Conditional outputs are not allowed. Each address of this ROM selects a word describing a STATE-QUALIFIER PAIR and thus this ROM is called a state-qualifier pair addressable ROM. Figure 5.13 shows the basic blocks for this type of ROM machine along with a typical state described by one ROM word. The state-qualifier pair ROM has some additional hardware called the SELECTOR and the SWITCH. The selector takes the test part of the ROM word and selects the corresponding qualifier as its output. The switch in turn selects one of the two link states on the basis of the logic level of the qualifier selected and presents this state as the next state for the state address register. The address corresponding to a test of 0 is called the DIRECT or FALSE (F) address, while the address corresponding to a test of 1 is called the ALTERNATE or TRUE (T) address. These two designations are also shown in Figure 5.13 by subscripts F and T on the link parts of the ROM word.

The address portion of a state-qualifier ROM is a simple complete decode of the n-input lines. The decoder is independent of the ASM chart description except for providing enough possible states. The entire description of the ASM chart is stored in the memory portion of the ROM in a state-by-state basis.

The ASM chart meeting the requirements for the state-qualifier ROM is stored in the ROM by a direct conversion from chart to word structure. For example, Figure 5.14 describes a portion of an ASM chart and the corresponding ROM word coding.

Assumed Addresses

Another ROM word structure uses only one link field while still using the state qualifier pair. The alternate address is stored in this field. The direct address is called an ASSUMED ADDRESS and is provided by adding 1 to the present address. This addition is usually accomplished by incrementing the address register although in Figure 5.15 this increment is indicated by a state (+) 1 selection by the switch portion of the circuitry external to the ROM. Figure 5.15 also shows a sample ASM chart state with two next states and a notation using a qualifier QX with only one next state. The second terminology represents a qualifier which is always 1 to force the state sequence to the true address stored in the ROM.

The assumed address ROM requires a forced true address to complete extra links caused by the limited scope of the assumed address. The extra links are called X-LINKS and are a characteristic of the ASM chart. For example, in Figure 5.16 heavy lines are drawn on the ASM chart to indicate assumed address transitions even before states are assigned. A group of assumed address transitions in a row form a LINK STRING. One link string begins in state (c) and goes to state (a), state (b), state (f), and finally to state (e). The jump from state (e) to state (a) must be an X-link address because each successive state in a link string has successive addresses, and since an assumed address is always an increment, it is never possible to jump to a previous state in the link string with an assumed jump. Therefore, there is a simple procedure for finding X-links in an ASM chart. First, find the smallest set of nonintersecting link-path strings that covers all the elements of the ASM chart description. (A decision box is covered when one of its link paths is included in a string.) Then, add an X-link at every point where, in order to cover an element, a jump is made to the middle of a link string. In Figure 5.16 there are two link strings and two X-links. In both of these X-links there is no qualifier on the state so no additional states are required. More often than not, additional states have to be added to the ASM chart to provide an X-link state. X-links are a penalty paid for saving a link field, and in algorithms which have many strings of operations, the penalty is small compared to the savings. Later in this chapter an example will demonstrate the assumed address more extensively.

Variable Format Addressable

A third basic word structure is called VARIABLE FORMAT. The link-path addressable ROM and the two-address ROM are FIXED FORMAT in comparison in that every bit of the ROM word has a single fixed function. In a variable format word the same bit may have several different functions depending on the position, address or other bits in the word. The basic variable format word has two formats determined by a single bit of the ROM word as shown in Figure 5.17. The first format is the instruction part of the ROM information. The second format is the link and the test part of the ROM information. A set of instruction, test and link parts takes two ROM words. The same information could have been stored in a single ROM word of twice as many bits, but often it is less expensive to limit the width of the ROM word. As in the previous ROM, the link part can be a two-address or an assumed address format. Figure 5.17 shows the block diagram for the variable format ROM and the ASM chart representation for the two formats. Format 1, selected by a 1 in the left most bit position of the ROM word, selects state output instructions, but since there is no link part an assumed address is used to select the next ROM word. The next ROM word may be either a format 1 or a format 2. A format 2 can make a test and jump to one of two next states. An assumed address requires that one of these next states be a successive state as shown. The penalty for a variable format ROM is that more state times are required to execute an algorithm than are required in a two-address fixed format ROM.

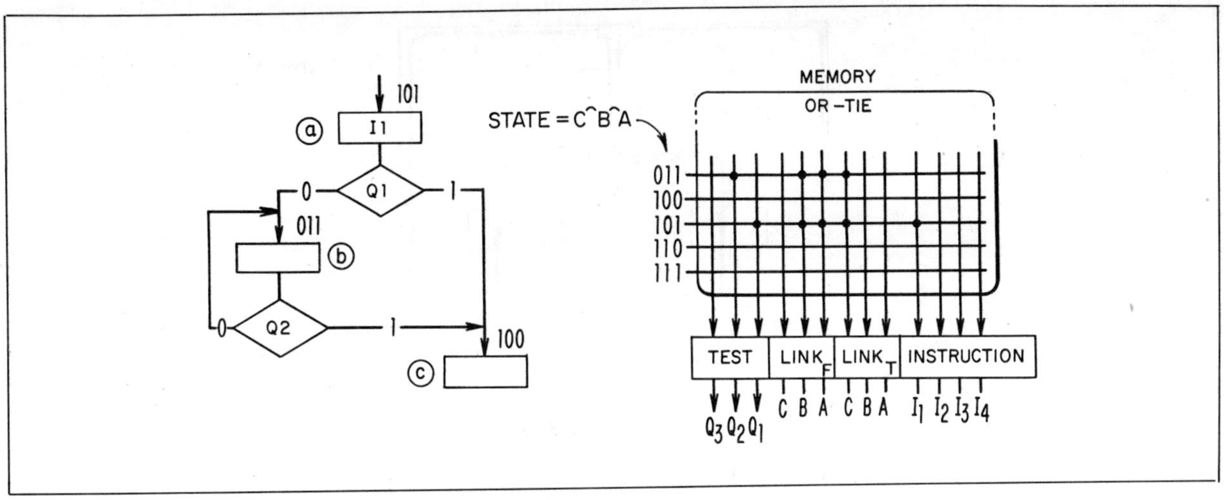

Figure 5.14 A Sample ROM Pattern for Two States.

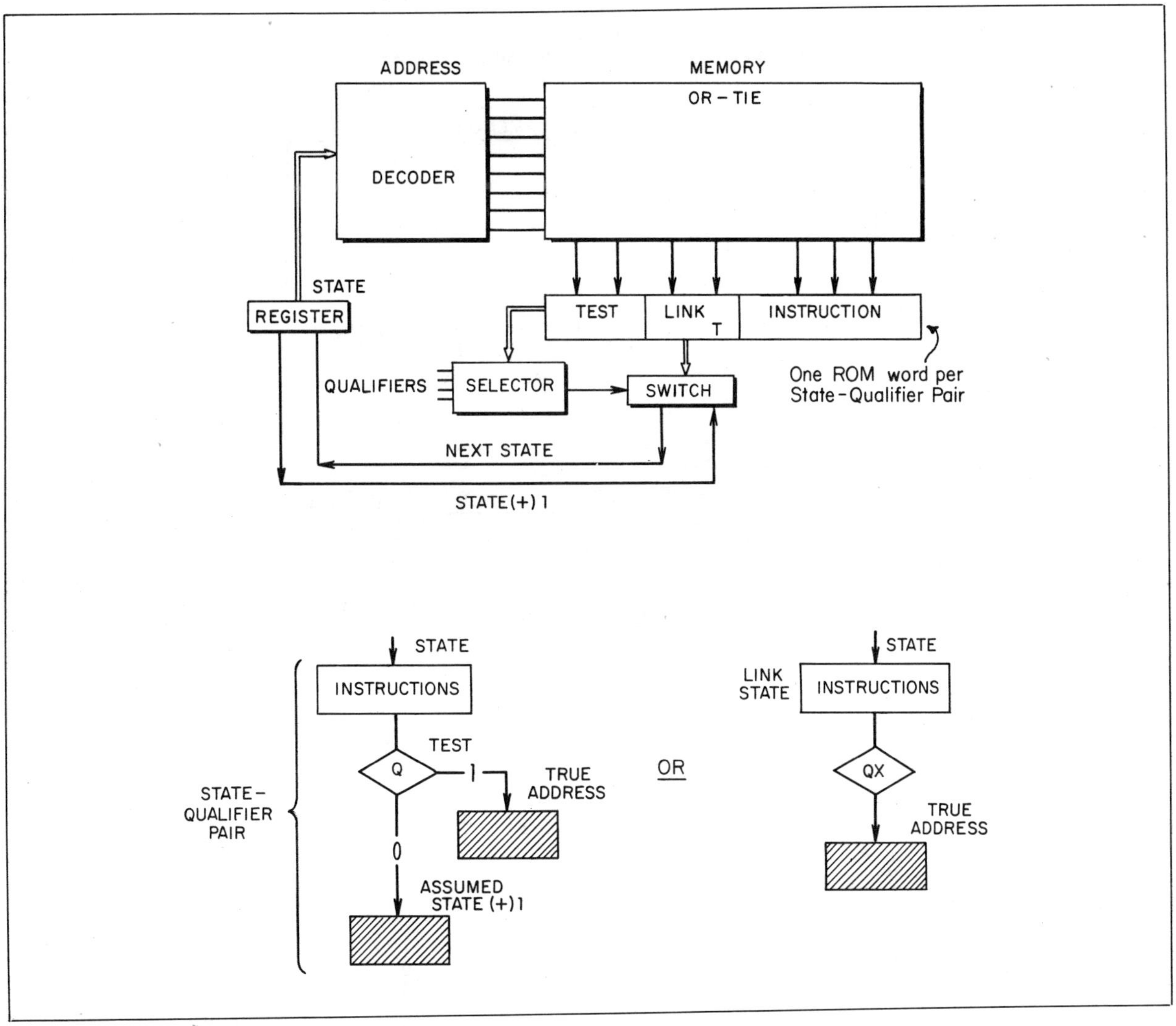

Figure 5.15 A ROM Structure Based on One Assumed Address, Fixed Format.

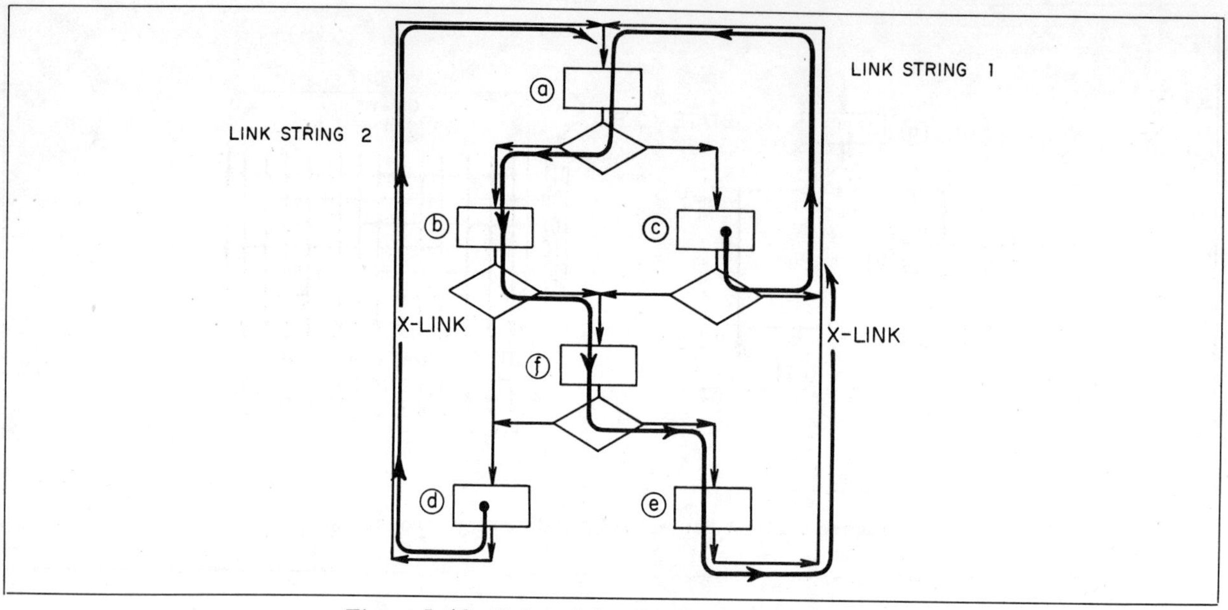

Figure 5.16 Determining the Number of X-Links by Forming Link-Path Strings in the ASM Chart.

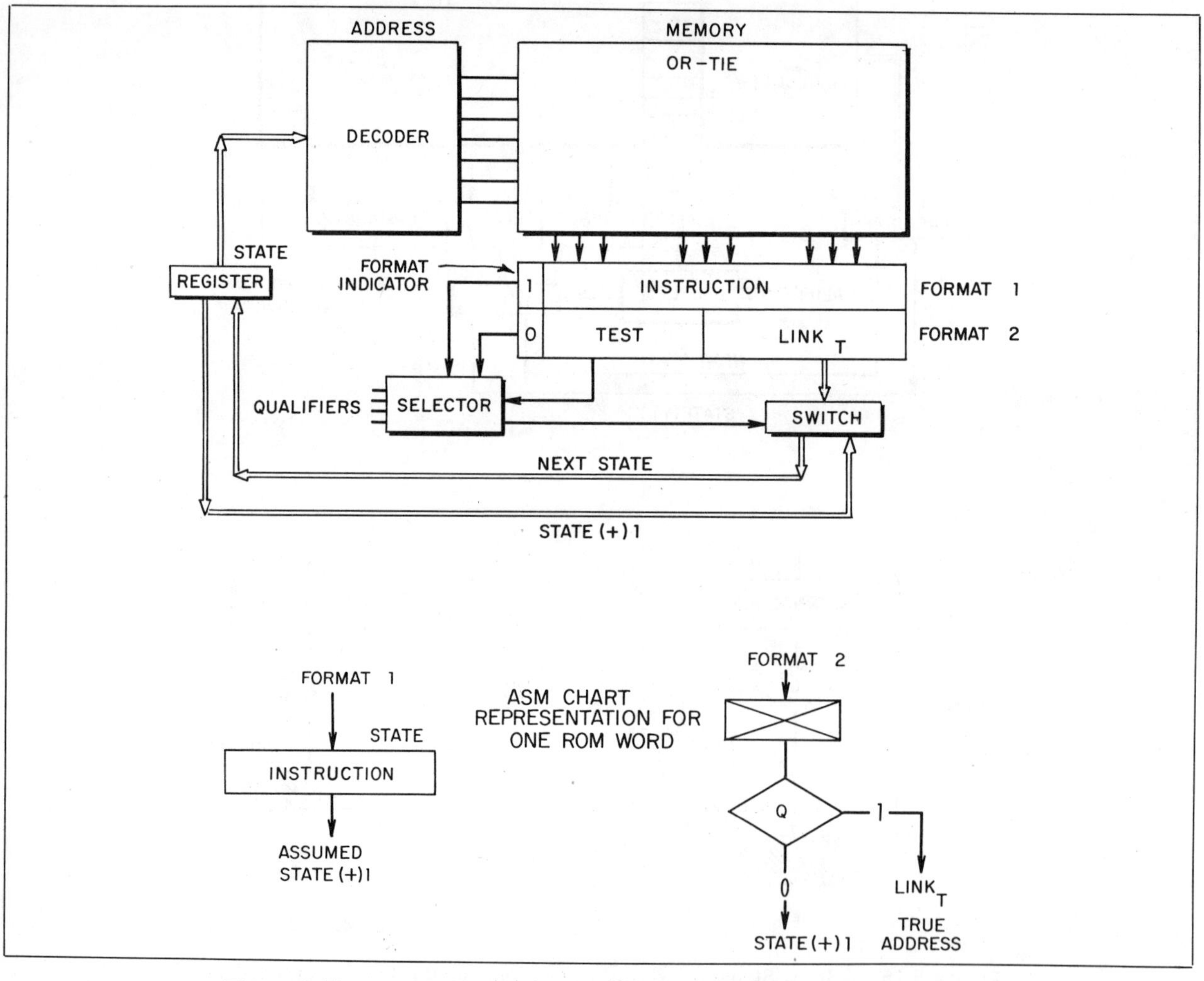

Figure 5.17 A Variable Format ROM Structure with an Assumed Address.

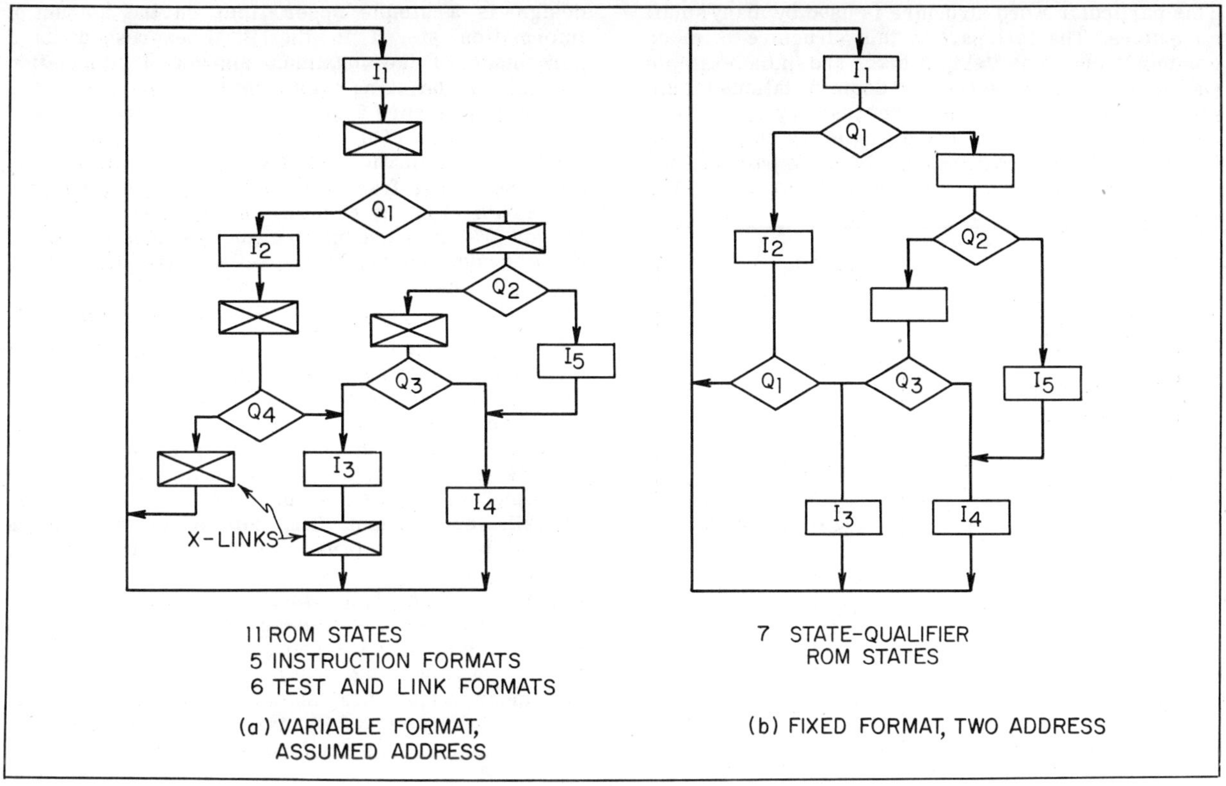

Figure 5.18 The Effect of Fixed and Variable Format ROM Word Structures on an Algorithm Description.

The comparison of state times and memory requirements for the variable and fixed ROM can be introduced by a comparison of the same algorithm done with both formats. The left ASM chart has two X-links and a total of 11 ROM states, while the right ASM chart has seven ROM states. The variable format ROM on the left, however, has a narrower word than the fixed format ROM on the right. If LI is the bit length of the instruction part and LL and LT are the corresponding lengths of the link and test parts, then if LL (+) LT >LI, 11 (LL (+) LT) is the number of bits required for the variable format ROM, while 7 (LI (+) LL (+) LT) is the number of bits required for the fixed format ROM. It is not clear which of these ROMs is smaller because it depends on the lengths of the particular parts of the ROM word. The next section is a detailed study of determining the required bit lengths for each ROM word part, but first several other word structures shall be briefly mentioned.

Other ROM Structures

The word structures discussed thus far are by no means all that are possible. Using the three basic elements of a word structure, the link part, the test part and the instruction part, a wide variety of structures can be created which will change the restrictions on the f and g functions in one way or another. For example, suppose in each state time two qualifiers were to be checked and four possible next states allowed. A word structure could be formed as follows which chooses one of four links on the basis of A and B as shown in Figure 5.19.

Test A	Test B	Link 1	Link 2	Link 3	Link 4	Instruction
		$A \cdot B$	$\bar{A} \cdot B$	$A \cdot \bar{B}$	$\bar{A} \cdot \bar{B}$	

Figure 5.19 A Four-Link Structure.

Suppose two possible instruction sets could be chosen by a test and each set had a different link. A corresponding variable format word structure might look like Figure 5.20.

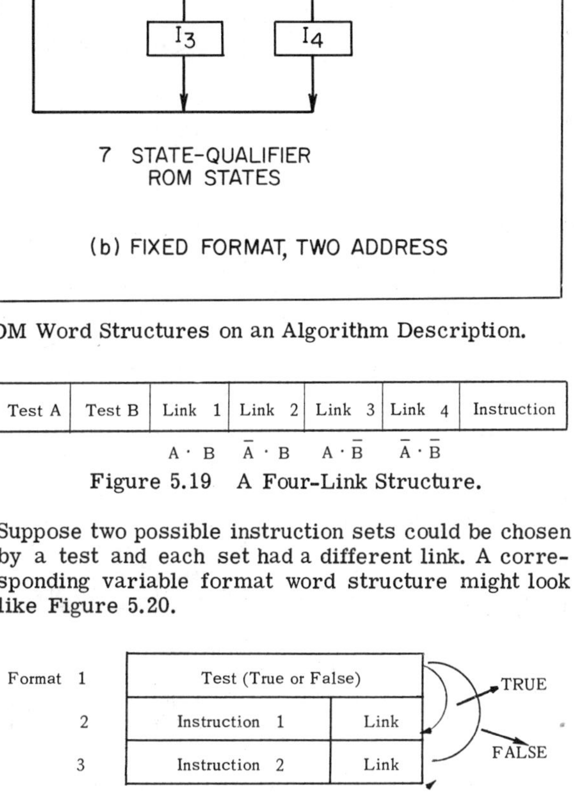

Figure 5.20 A Three-Format Structure.

Suppose the ROM was made very narrow and time could be used. Then the three parts could be split as shown in Figure 5.21.

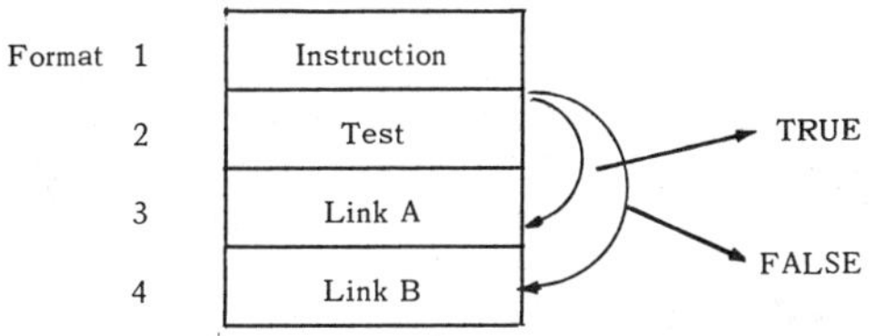

Figure 5.21 A Completely Separated Word Structure.

This particular word structure is used by many small computers. The test part in this structure is more commonly known as "skip if test" and in the example shown is a "skip if test false." Some variations on the structure shown are possible because any format may be used in any position. For example, a string of tests can be made or a wait loop can be made by combinations of word formats as shown in Figures 5.22 and 5.23.

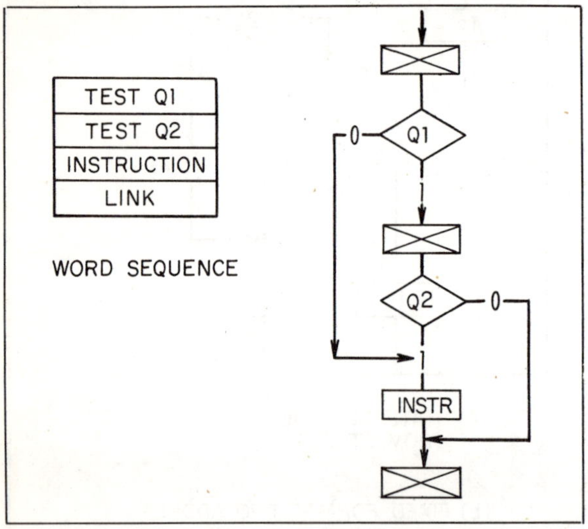

Figure 5.22 A Multiple Test Sequence.

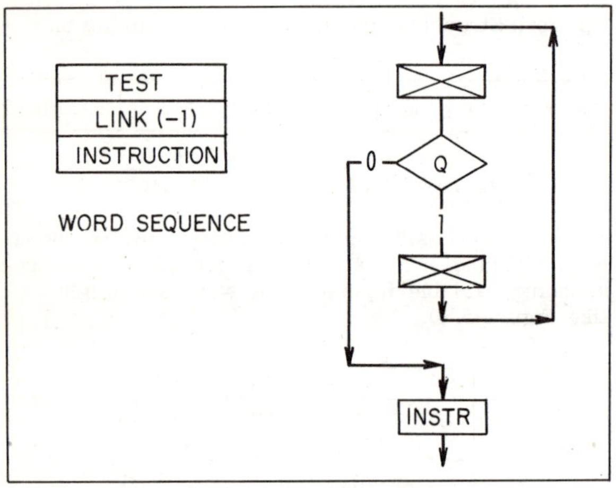

Figure 5.23 A Wait Loop Sequence.

5.3 INFORMATION STORAGE REDUCTION

The information stored in tne OR-TIE array of a ROM machine may require fewer bits of storage than at first seems required. A ROM pattern may be reduced if the coding efficiency is low. Some methods of reducing ROM storage are described in this section. The new concepts introduced are the measurement of CODING EFFICIENCY, the use of BIT PACKING, and the use of FUNCTION EXTRACTION.

Coding Efficiency

The value of coding efficiency (CE) used in ROM design is a simple upper limit on the amount of information stored in the ROM expressed as a percentage of the maximum amount of information storable in the same number of bits. The percentage is plotted in Figure 5.24 as a function of the fractional bit content. (This curve is derived shortly but first the application of the value of CE(max) is discussed.) The fractional bit content is obtained for any given ROM by counting the smaller number of 1's or 0's in each bit line and dividing the sum of these counts by the bit capacity of the ROM. When each bit line has fewer 1's than 0's, the fractional bit content is just the fraction of 1's in the ROM.

CE(max) is used as a necessary but not sufficient condition for a good ROM coding. It is necessary to have a high CE(max) (80% or greater) to have a reasonable ROM coding; however, a high CE(max) does not insure that this ROM coding has been achieved. For example, Figure 5.26 shows an OR-TIE array for which the smaller number of 1's or 0's is listed opposite each output. The sum of these numbers divided by the total number of bits (bit positions) in the ROM is 0.321 as shown. This number is the fractional bit content which corresponds to a CE (max) of about 91%. This means that as little as 9% of the ROM bits may not convey information. Probably CE is less than the maximum but it is not clear how much. Figure 5.25 shows another OR-TIE array where it is clear that CE(max) gives a false indication of the actual amount of information contained.

The factor which is left out of CE(max) is the degree of randomness of the bit pattern in the array. If randomness is high then the actual CE approaches CE(max). Although it is difficult to mathematically determine the randomness of an array, the eye does a fair job. A random array of bits will have a lack of any pattern in the bits somewhat like the array in Figure 5.26 as compared to Figure 5.25 which has a simple, clear pattern.

The derivation of CE(max) is based upon the equation from information theory which describes the number of bits of information, N, contained in a bit string n bits long given the probability of finding a 0, P_0, and the probability of finding a 1, P_1, as follows.

$$N = \left[P_0 \log_2 \frac{1}{P_0} + P_1 \log_2 \frac{1}{P_1}\right] n \qquad (3)$$

The coding efficiency is defined as:

$$CE = \frac{N}{n} 100\% = \left[P_0 \log_2 \frac{1}{P_0} + P_1 \log_2 \frac{1}{P_1}\right] 100\% \qquad (4)$$

The probability of $P_0 = 1-P_1$ so CE may be expressed in terms of P_0 or P_1. The maximum of CE occurs when $P_0 = P_1 = 0.5$ as follows:

$$CE \Big|_{P_0 = 0.5,\ P_1 = 0.5} = \left[0.5 \log_2 2 + 0.5 \log_2 2\right] 100\% = 100\%$$

The value of P_1 can be estimated by the number of 1's in the string if the 1's are randomly distributed taken as a fraction of the total bit string. This value will be the maximum value for P_1 and therefore provides a convenient means of calculating CE(max). Since any output may be complimented without changing the essential nature of the information stored in the ROM and since CE is symmetrical about $P_0 = P_1 = 0.5$ then the sum of the lowest count of 1's or 0's in each output line is a measure of P_0 or P_1 in terms of giving a value of CE(max). This number is called the FRACTIONAL BIT COUNT. The equation for CE(max) is plotted in Figure 5.24 for one half of the total probability range.

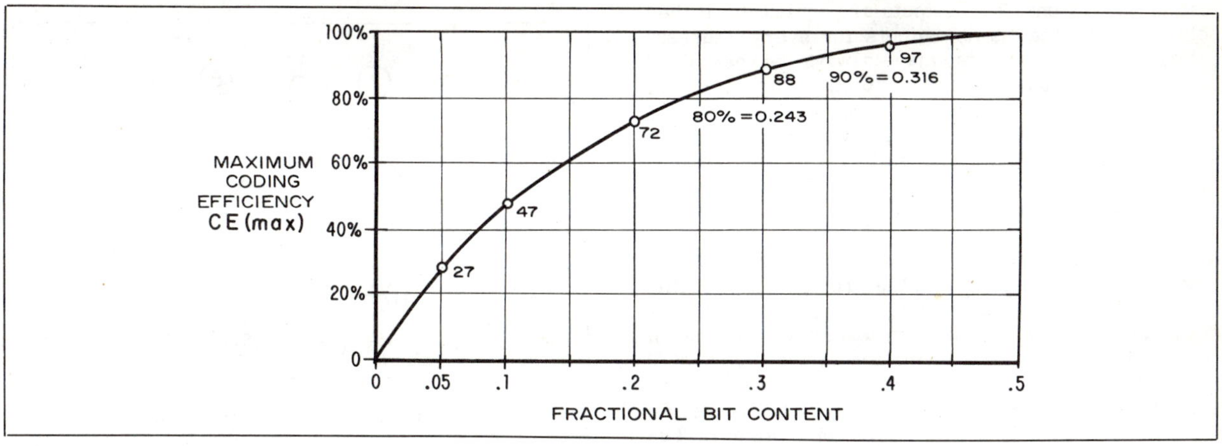

Figure 5.24 Maximum Coding Efficiency for Memories.

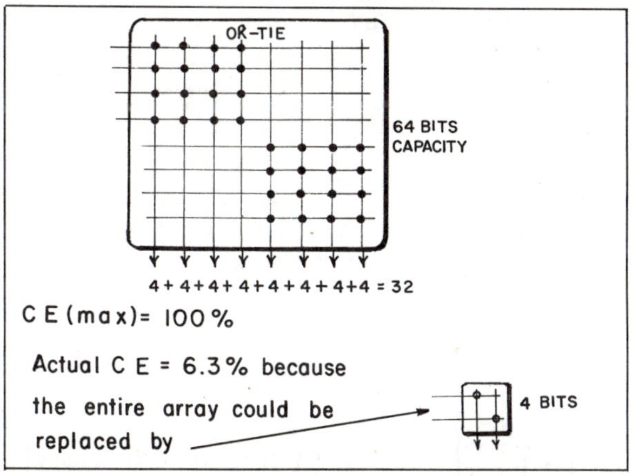

Figure 5.25 An Example Illustrating the Large Discrepancies That Can Exist between the Actual and the Maximum Coding Efficiency.

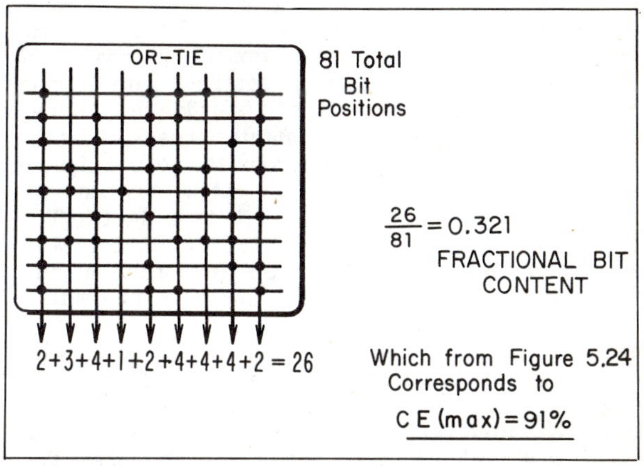

Figure 5.26 An Example Calculation of CE(max).

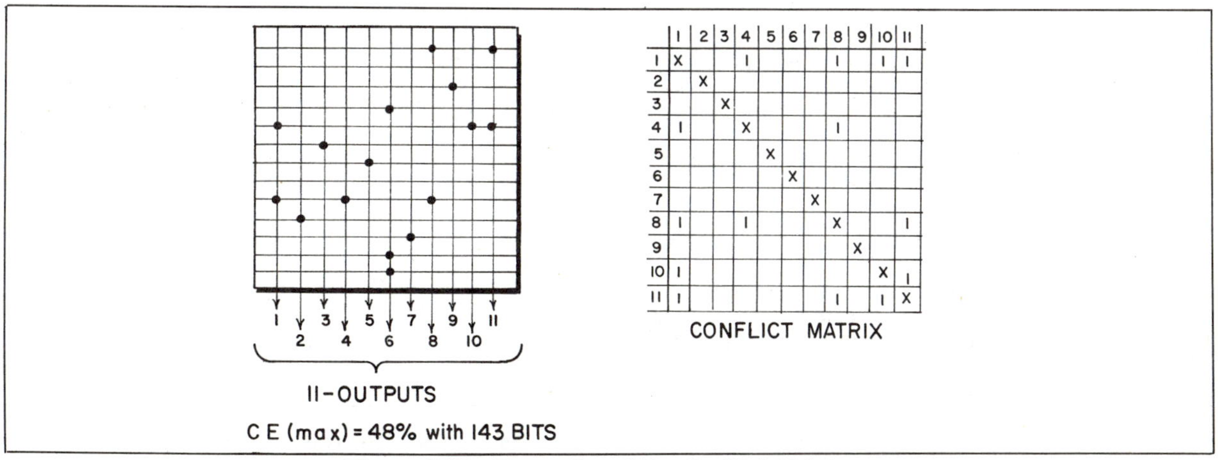

Figure 5.27 Formation of the Conflict Matrix.

Bit Packing

When CE(max) is very low the bits in the ROM may be condensed into a smaller ROM. The techniques covered here involve reducing the width of the ROM word by combining or eliminating output lines. EQUIVALENT output lines may always be combined by eliminating all but one. Two outputs are equivalent if they both produce the same outputs for every address. COMPLEMENT outputs may always be combined by eliminating one line and complementing the other. Two outputs are complements if they produce different outputs for every address. Exclusive groups of outputs may be replaced by fewer lines by coding the outputs. Exclusive outputs are never 1 at the same time. Such outputs warrant further discussion.

Although many exclusive groups of outputs can be determined by the nature of the problem, i.e., arithmetic functions, input-output buses, etc., the task of identifying the remaining exclusive groups of outputs is aided by the formation of the CONFLICT MATRIX. The conflict matrix lists all the outputs that conflict in pairs. For example, Figure 5.27 gives an OR-TIE array with a low value of CE(max). The conflict matrix is formed by comparing the outputs two at a time. An output compared to itself is indicated by X's along the diagonal. The conflict matrix is symmetrical about this diagonal. In row 1 of the conflict matrix, output 1 conflicts with outputs 4, 8, 10, and 11 as entered by 1's under the corresponding column. These same entries appear along the 1's column opposite the appropriate row. Output 4 conflicts with output 8, 8 with output 11, and 10 with 11. There are no other conflicts in the OR-TIE OUTPUTS. From the table a CONFLICT DIAGRAM is constructed by a special representation of the outputs with interconnecting lines representing each conflict as shown in Figure 5.28. In this array a choice of an exclusive group of outputs corresponds to a group with no conflicts. If output 8 were chosen for the group, then output 4, 11, and 1 could not be chosen but output 10 could be. However, if output 1 were chosen for the group, then neither 4, 8, 10, nor 11 could be chosen. The largest exclusive group for this array is found by choosing the outputs with the fewest conflicts first, which leads to the following group:

$$\{2, 3, 5, 6, 7, 9, 4, 11\}$$

This group has eight elements. At first it may appear that these elements could be encoded in 3 bits. However, one code must be reserved to indicate no outputs. Therefore, an n-bit code can only code $2^n - 1$ outputs. It is most desirable to find groups of $2^n - 1$ outputs so that all the codes can be used. In the above group, output 2 will be left out of the exclusive group. The remaining outputs have a conflict diagram as shown in Figure 5.29 and may be grouped into three outputs and one output as is shown. The net result of finding the exclusive groups in the ROM outputs is the packed ROM shown with decoders in Figure 5.30. The ROM and decoders produce the same outputs as before, but now the value of CE(max) is almost doubled and the number of ROM bits required is almost halved. The manual use of the conflict matrix and the conflict diagram is useful for small problems. On larger problems this same procedure has been computerized with gratifying results. A problem using a ROM of 43 addresses and 29 outputs was reduced to 19 outputs in less than 2 minutes. This example leaves out many details in the full algorithm for searching to find the best set of exclusive groups, but it should serve to indicate the procedure and the effectiveness.

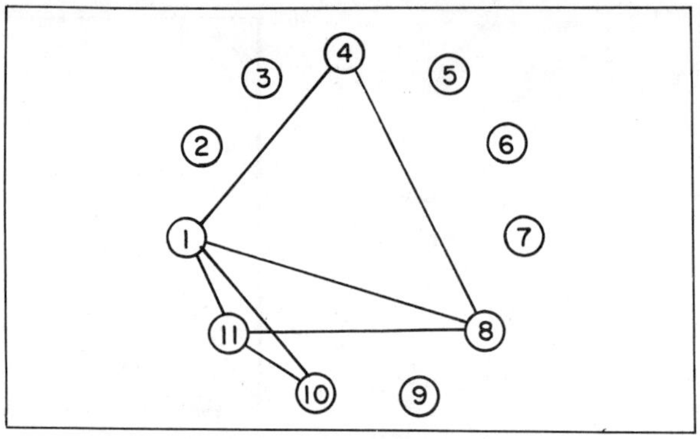

Figure 5.28 A Conflict Diagram.

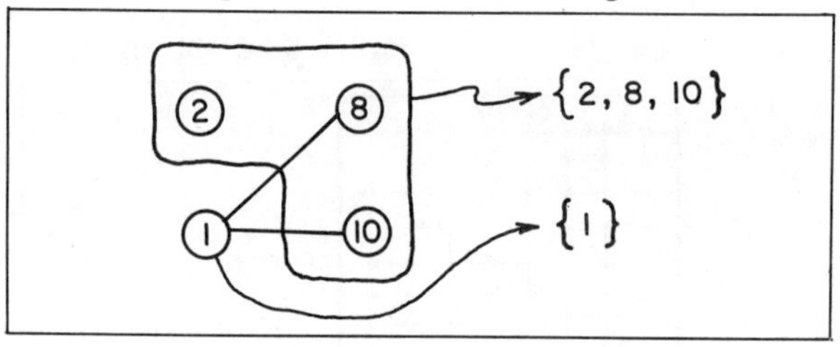

Figure 5.29 The Remaining Outputs after the First Reduction of Figure 5.28

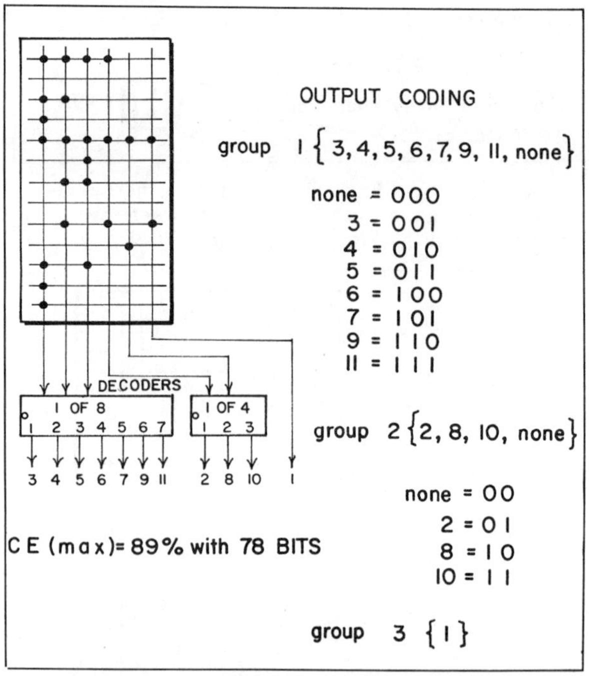

Figure 5.30 The Compacted Version of the ROM Shown in Figure 5.27.

Function Extraction

Common functions between several ROM bits may be used to reduce the number of required ROM bits. Two or more outputs may contain the information to generate an output by forming simple AND or OR combinations between them. Some of these relationships may be established when the instruction set is formed. A common example is two outputs which generate a third output when neither of the first two are present. Only two ROM lines are required, with a NOR generating the third output.

Although function extraction can reduce the number of lines, it is only most effective when don't care outputs are used to increase the chances of finding common functions. Don't care outputs occur for any state where the output does not affect the validity of the algorithm. Finding don't care outputs involves considerable work since each output must be checked in each state. Where reducing the ROM is very important, however, the extra work required to isolate the don't care outputs may be rewarded by some additional reduction by common functions once the exclusive groups are extracted by bit packing. The reduction in ROM, of course, is accompanied by the addition of external gating.

Changeability and Compaction

The basic ROM with one output per ROM bit has complete flexibility in choice of outputs in each state. The storage reduction by coding exclusive output groups places restrictions on which instructions can be given simultaneously in any state. The reduction by common functions places the particular function restrictions on the simultaneous issuing of certain instructions, which tends to be more confusing than the simple exclusive groups. The result is that the more a ROM is compacted the more restrictive it becomes in allowing a change without completely reorganizing the ROM packing.

Even though common functions can reduce a ROM, the extraction of exclusive groups does not involve finding don't cares and only restricts the instructions into simple exclusive groups. Also, the search for exclusive groups can be executed with enough ease that even conflicting changes can probably be reworked into another possible grouping without undue time spent in the process.

5.4 COMPARATIVE IMPLEMENTATIONS

This section implements the Black Jack control described in Sections 3.7 and 3.8 with a two-address, fixed format ROM and with an assumed address, variable format ROM. Some general guidelines for choosing a ROM word are given in conclusion.

Two-Address, Fixed Format

The basic Black Jack control algorithm written in terms of state qualifier pairs is given in Figure 5.31. It is assumed that the ROM implementation must be fully synchronous because there are unknown race-producing delays possible in the ROM decoding. For this simple example, *YCRD is transformed to a synchronous qualifier by running it through a clocked delay flip-flop. (See Section 4.7.) There are better ways to handle many asynchronous qualifiers which are described in Chapter 6. The additional states required to handle the multiple qualifier in state (d) make a total of 11 state-qualifier pairs.

A ROM word is determined by finding the number of bits required for the link, test and instruction parts of the ROM. There are 11 state-qualifier pairs so 4 bits are required to code the ROM states. There are five qualifiers plus one for no qualifiers making a total of six test conditions so 3 bits will code the test part. There are 11 instructions but seven of them can be grouped into an exclusive set requiring only 3 bits. These bits when added to the 2 bits required for the remaining three instructions make a total of 5 bits for the instruction part. The formation and coding of the ROM word chosen are summarized in Figure 5.32.

An even better compaction of ROM bits for the instruction part is possible by the extraction of common functions, which requires only 3 bits of coding as shown in Figure 5.33.

This new instruction coding makes a total ROM word of 14 bits. Now, since the state assignment is completely flexible, the last 3 bits of the state can be made the same as the codes in Figure 5.33. Using a 1 out of 10 decoder with an inhibit or 4th bit line which ignores all codes greater than 10, the other states will generate no output at all if used as the input directly. This simplification eliminates the instruction part of the ROM completely making only 11 bits in the ROM word. A possible state assignment for this simplification is given in Figure 5.31.

Using this assignment and the codes for the test part, a tabular description of the ROM information may be constructed as shown in Figure 5.34, with the logic circuit results as shown in Figure 5.35.

Assumed Address, Variable Format

The ASM chart for a variable format ROM implementation with the test and link parts in one format and the instruction part in another involves adding extra states to define the additional formats. A modified ASM chart of the Black Jack control with X-links added where required to meet the sequence requirements is shown in Figure 5.36.

The instructions IHBRK, IHSTND and IHHIT must be modified because with this type of ROM it is only possible to give an instruction intermittently while checking a qualifier. As shown by states (g), (q), and (p) the instruction IHBRK is issued once every three clock times while YCRD = 0. The same is true of IHSTD in state (h) and IHHIT in state (k). The restrictions imposed by an assumed address have raised the total number of states to 20.

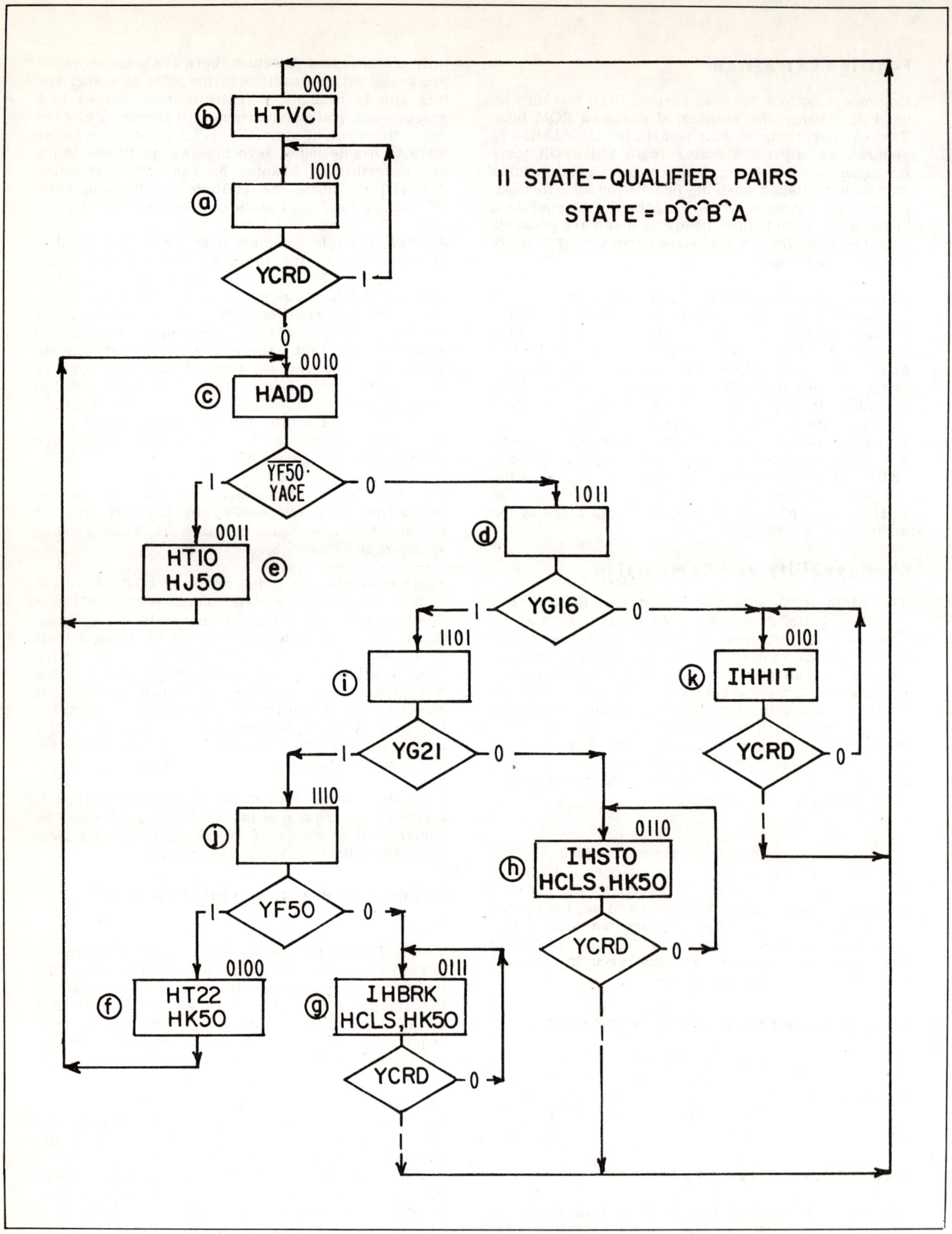

Figure 5.31 State-Qualifier Pair Notation for the Black Jack Control.

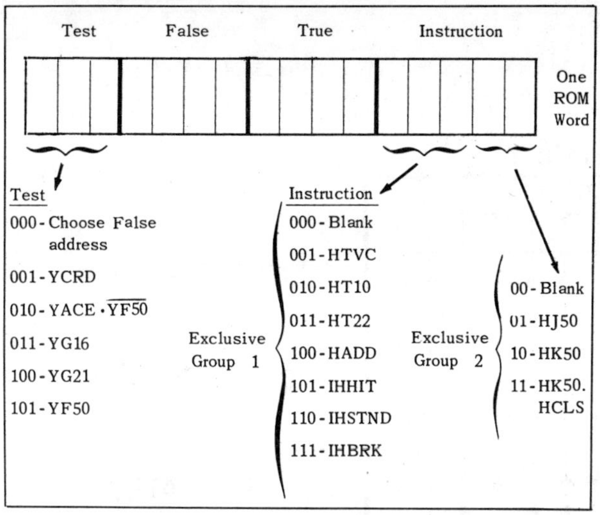

Figure 5.32 Structure and Partial Coding for a Two-Address ROM Word.

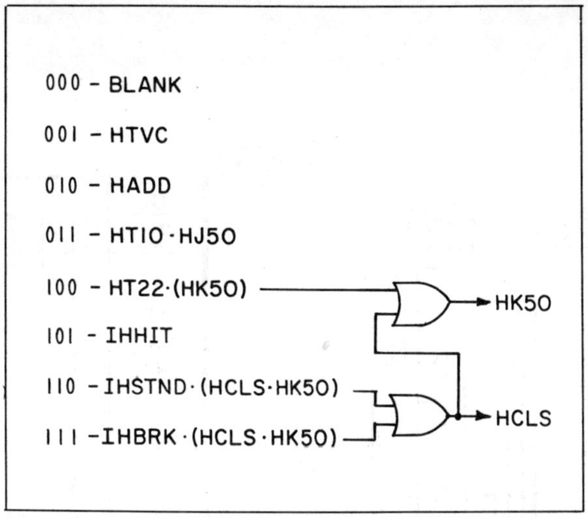

Figure 5.33 An Improved Coding for the ROM Word.

SYMBOLIC TABLE				ROM CODE TABLE		
State	Test	False	True	Test $T_3T_2T_1$	False D C B A	True D C B A
a	YCRD	c	a	001	0010	1010
b	FALSE	a	-	000	1010	—
c	$\overline{YF50}$· YACE	d	e	010	1011	0011
d	YG16	k	i	011	0101	1101
e	FALSE	c	-	000	0010	—
f	FALSE	c	-	000	0010	—
g	YCRD	g	b	001	0111	0001
h	YCRD	h	b	001	0110	0001
i	YG21	h	j	100	0110	1110
j	YF50	g	f	101	0111	0'00
k	YCRD	k	b	001	0101	0001

Figure 5.34 A Table Describing the ROM Code Pattern for a Two-Address Organization.

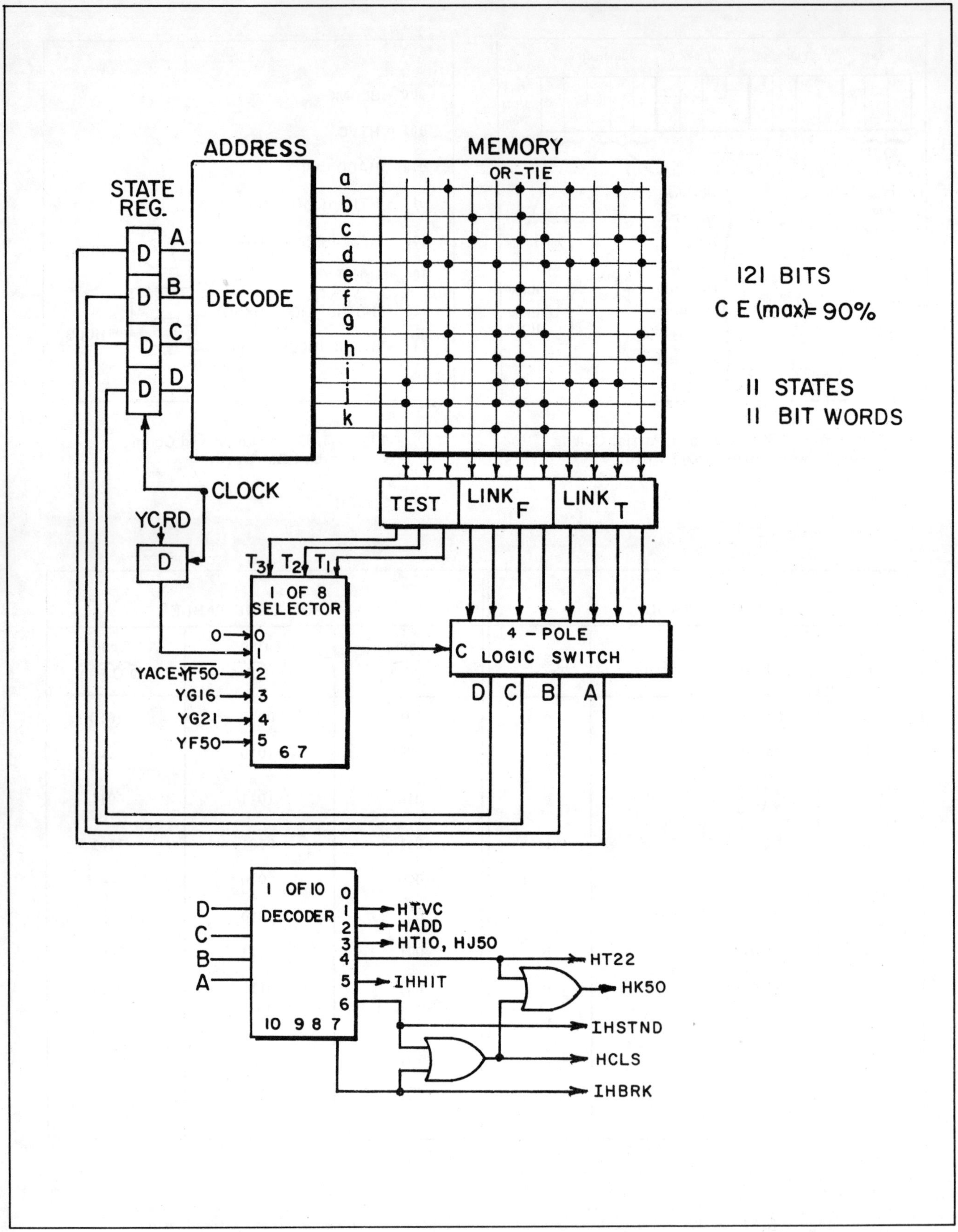

Figure 5.35 ROM Implementation of the Black Jack Control Using a Two-Address ROM.

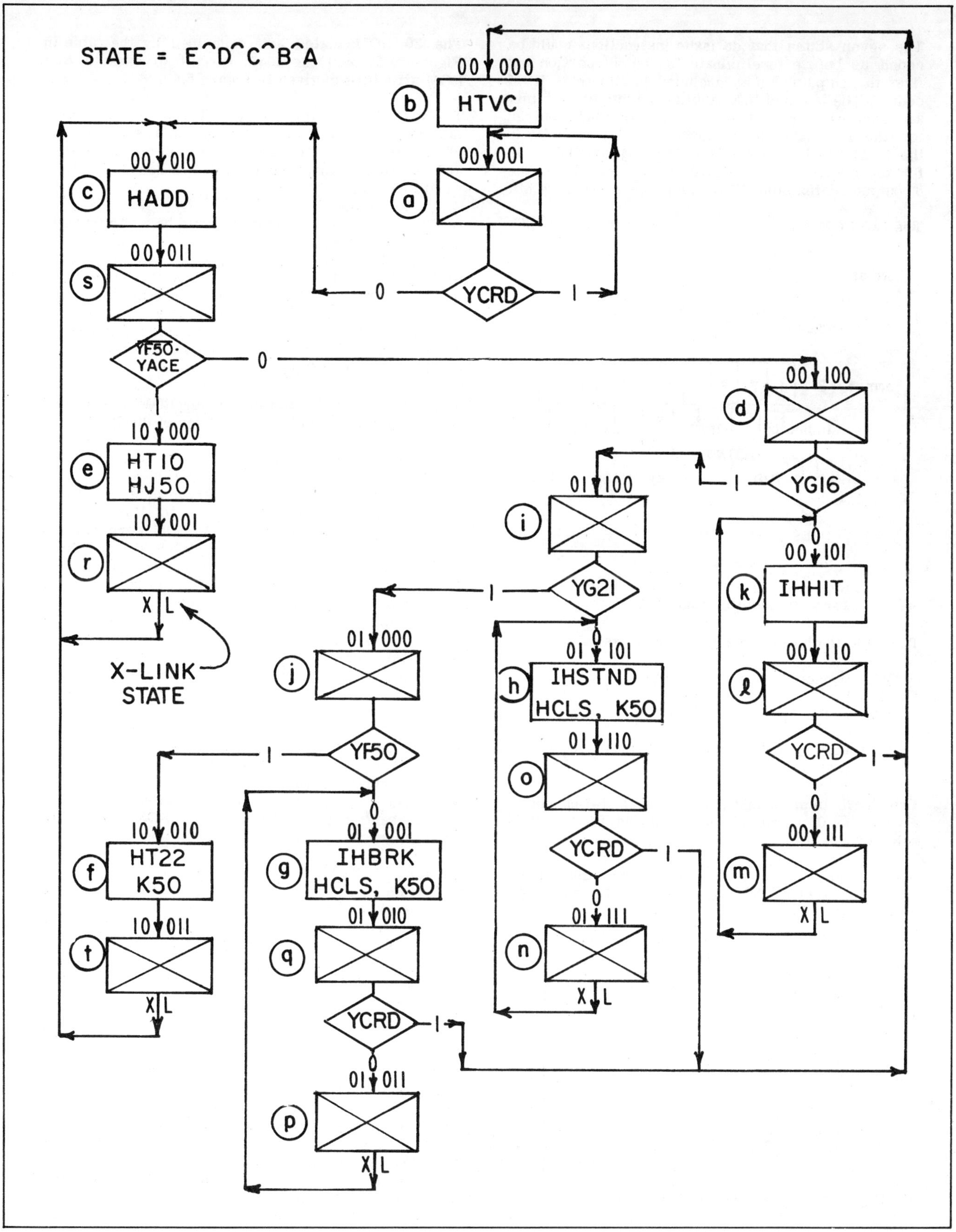

Figure 5.36 Assumed Address, Variable Format ASM Chart Description of the Black Jack Control.

The seven states that do issue instructions could be coded as before to eliminate the output portion but first the outputs will be included to illustrate ROM coding. The test and link parts take 8 bits since 5 bits are required for the link. One additional bit required for the format indicator makes a total of 9 bits in the ROM word. The 8 bits of the ROM are available for the outputs so one bit will be used for each of the 8 output states and HCLS is given a separate bit.

The two ROM formats appear as shown in Figure 5.37.

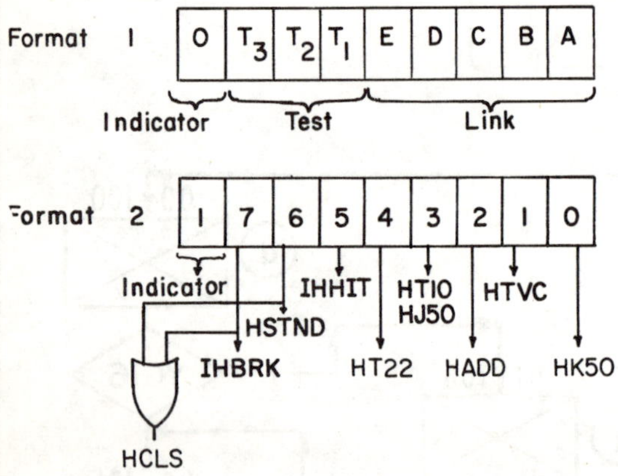

Figure 5.37 The Two ROM Formats for a Variable Format ROM.

The logic to choose the assumed address is:

YAA = choose assumed address

YAA = FMT1 · TEST + FMT2

$$\text{YAA} = \overline{\text{INDICATOR}} \cdot \text{TEST} + \text{INDICATOR}$$
$$= \text{TEST} + \text{INDICATOR}$$

This logic implies that the test for an X-link should always be 0, which is easily chosen by the TEST selector.

A state assignment is made by choosing link strings of states. Figure 5.36 is given a possible assignment which is shown on a map in Figure 5.38.

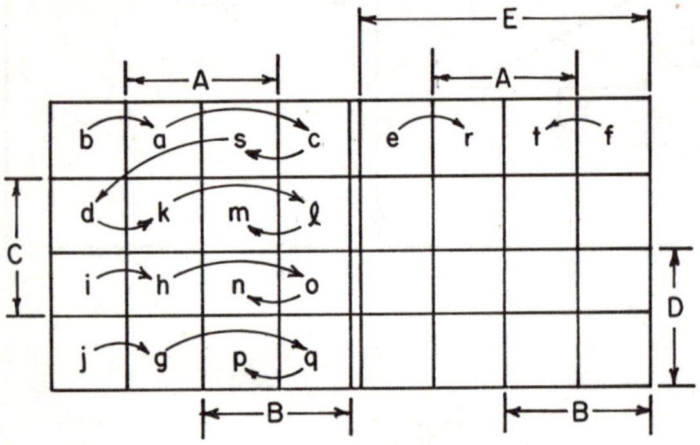

Figure 5.38 A Possible State Assignment for the Assumed Address ROM. (Arrows show assumed transitions.)

The 20 ROM states are described by the table in Figure 5.39, and the logic circuit to execute this ROM structure is given in Figure 5.40.

The variable format implementation is more complex than the fixed format implementation. The added complexity is due to states added to complete the X-links and additional hardware required to decode the two-word formats.

Neither ROM implementation would be a particularly good replacement for the gate implementation in Chapter 4 because the ROM is so small that the penalties in added circuits cancel the advantages of the symmetrical array. This problem illustrates the cost problem with ROMs. The cost per bit increases with a decrease in ROM size until at some point, about 50 states, gates generally become more economical than ROMs.

There is a generalization concerning ROM capacity and word width which aids in the choice of the ROM word width for any machine. A ROM organized with one state per algorithm state has the fewest number of states possible for that algortihm and, for a given clock, will be the fastest machine. To improve upon this time, the instruction set must be changed to allow reducing the number of states. But, if the ROM word width is restricted, the number of ROM states will increase, which corresponds to a slower execution time. However, the ROM capacity measured by the number of words times the word width has a minimum for a given restriction on ROM word width; and since the cost of the ROM in general corresponds to the capacity, this minimum corresponds to a minimum cost. This relationship may be illustrated by considering the Black Jack control as described in Chapter 4. This description has a minimum number of states for the instruction set. A ROM with the same number of states would have to have the capacity to store five next-state links in order to handle the links from state ⓓ. This ROM would require a 196-bit capacity. By restricting the word width to include only two next states as was done in this section for the fixed format ROM, five states were added to cover the transition and state output restrictions. The resultant 11-state ROM has a capacity of 121 bits. The word restriction to decrease the width to 9 bits adds nine more states to the ROM and the capacity of the ROM increased again to 180 bits. The relationship between ROM word width and the ROM capacity is generalized as shown in Figure 5.41.

This graph indicates that a word width can be chosen which minimizes the ROM capacity at some sacrifice to the machine speed.

The best word width can be estimated by finding the length of the link, test, and instruction parts. The link part consists of a number of addresses which may be estimated by the average number of links per state in the reduced algorithm description. The test part and instruction part are determined by the coding required for these sections. For example, in the Black Jack machine described in Chapter 4, the average number of links per state is 2.2, which indicates that the two-address ROM is a good word choice as is observed by its small capacity.

	SYMBOLIC TABLE			ROM CODE FORMAT 1		ROM CODE FORMAT 2
STATE	TEST	LINK	OUTPUT	TEST $T_3 T_2 T_1$	LINK E D C B A	OUTPUT 7 6 5 4 3 2 1 0
a	YCRD	a	—	001	00 001	—
b	—	—	HTVC	—	→	00000010
c	—	—	HADD	—	→	00000100
d	YG16	i	—	011	01 100	—
e	—	—	HT10 HJ 50	—	→	00001000
f	—	—	HT22 HK50	—	→	00010001
g	—	—	IHBRK HCLS, HK50	—	→	10000001
h	—	—	IHSTND HCLS, HK50	—	→	01000001
i	YG21	j	—	100	01 000	—
j	YF50	f	—	101	10 010	—
k	—	—	IHHIT	—	→	00100000
l	YCRD	b	—	001	00 000	—
m	X-L	k	—	000	00 101	—
n	X-L	h	—	000	01 111	—
o	YCRD	b	—	001	00 000	—
p	X-L	g	—	000	01 011	—
q	YCRD	b	—	001	00 000	—
r	X-L	c	—	000	00 010	—
s	YF50• YACE	e	—	010	10 000	—
t	X-L	c	—	000	00 010	—

5.39 Tabular Description of the Assumed Address, Variable Format ROM.

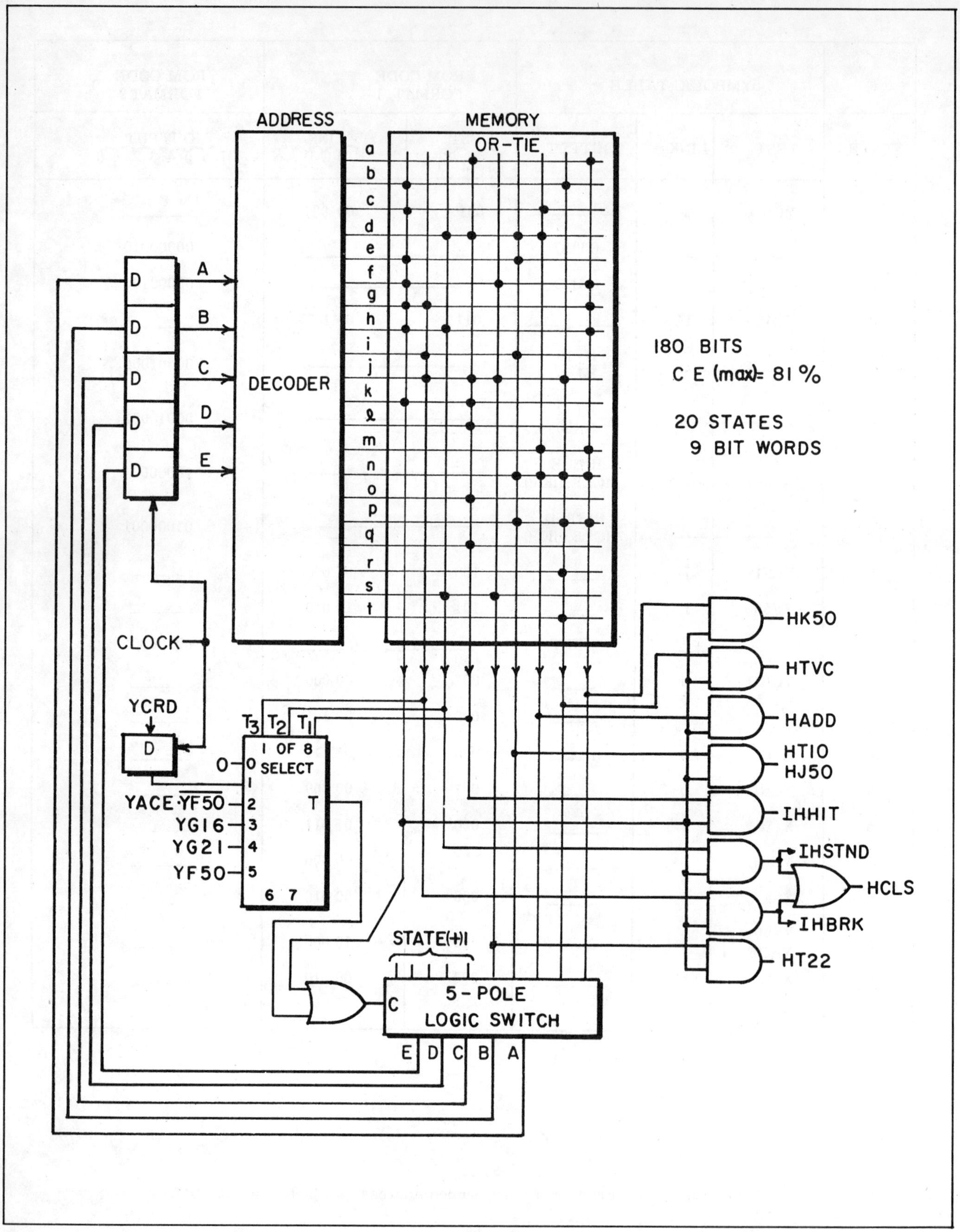

5.40 An Assumed Address, Variable Format Implementation of the Black Jack Control.

It is seldom possible in complex machines where ROMs are often used to construct the reduced algorithm to determine the average number of links per state. In these situations, the ROM structure must be based upon an estimation of the algorithm. The number of links per state will in general be close to 1 for arithmetic algorithms, close to 2 for control operations and possibly but rarely as high as 3 or 4 in very complex decision logic.

In many designs the controlling factor in word width is that the ROMs come with 2^N words and 1, 2, 4 or 8 bit widths. This fact alone may make a narrower word more optimum than the ideal because the additional words may be available, or a wider word may be used because extra bits are left after the required number are provided.

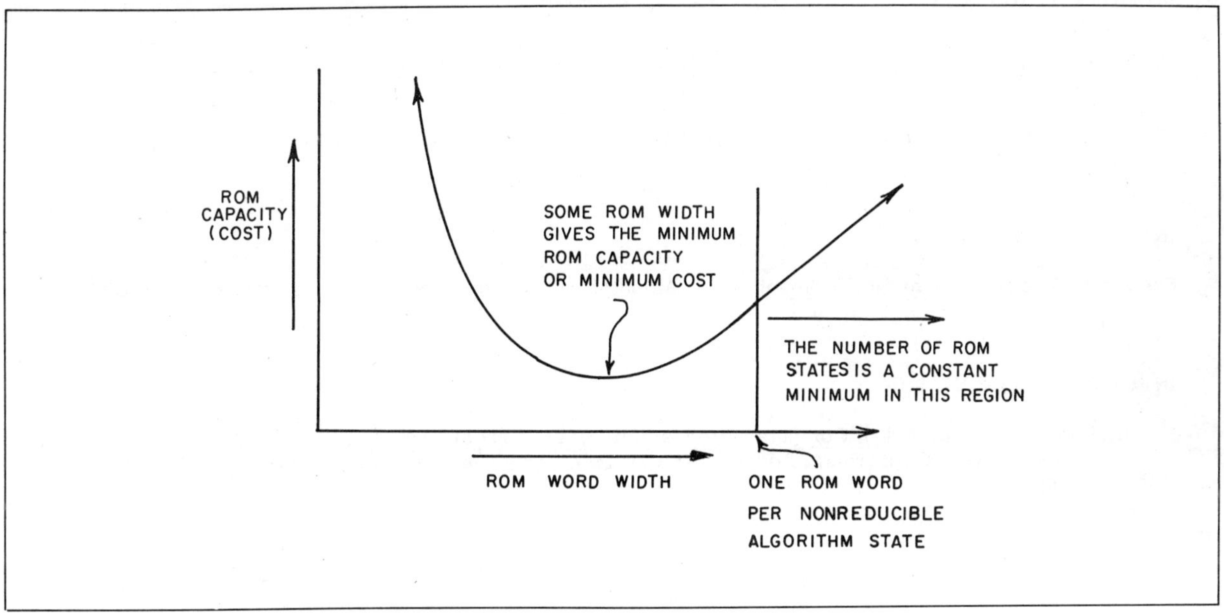

Figure 5.41 The Generalized Relationship between the ROM Word Width and the ROM Capacity or Cost for a Given Algorithm.

REFERENCES
CHAPTER V

MEMORY STRUCTURES

Hellerman, H.: Digital Computer Systems Principles (New York: McGraw-Hill Book Company, 1967), Chapters 3, 8.

Phister, Montgomery, Jr.: Logical Design of Digital Computer (New York: John Wiley & Sons, Inc., 1966), Chapter 7.

INFORMATION STORAGE

Schwartz, Misha: Information Transmission, Modulation and Noise (New York: McGraw-Hill Book Company, 1959), Chapter 1.

INFORMATION STRUCTURES

Knuth, Donald E.: The Art of Computer Programming, Vol. 1: Fundamental Algorithms (Menlo Park, California: Addison-Wesley Publishing Company, Inc., 1968), Chapter 2.

CHAPTER VI
LINKED STATE MACHINES

6.0 LINKED STATE MACHINES

The previous two chapters have described the design of algorithmic state machines. This chapter discusses the effects of several state machines working together. These state machines are said to be LINKED. The general linking of many machines is called ARRAY LINKING, but the coverage of this topic is beyond the scope of this book. Instead, four special facets of linking are discussed: INTERFACE LINKING, ITERATIVE LINKING, INTERPRETIVE LINKING and SOFTWARE LINKING. The significance of this chapter lies in specifying the relationship between logic systems, machine organization and software.

6.1 INTERFACE LINKING

Machines that are linked by only a few states are called INTERFACE LINKED. Examples of such linkages are the connections to computers of such instruments as line printers, teletypes, digital counters or A/D converters. Each of these instruments has a number of states just as the computer has a large number of states. The computer interfaces with the device in a relatively few number of these total states.

Serial Linking

Two machines that are interface linked in time, such that one or the other is primarily active at any one time, are called SERIALLY LINKED to describe their time-serial behavior. An example of two serially linked machines is given in Figure 6.1. Just the states of each machine relevant to the linking process are shown. States ② and ③ are named the CALLING STATES in the sense that one machine calls on the other to perform. The IB instruction from state ② causes machine B to leave state ③ while machine A waits in state ① for the IA call from machine B. This example assumes that both machines use the same clock.

If machine B had a shorter state time than machine A, the linking states would have to be arranged as in Figure 6.2. The additional complexity is required to ensure that the IA instruction is detected by machine A during its stable period. Machine A waits for IA.

When the relative state times are unknown, the ASM chart solution for serially linking two machines is given in Figure 6.3. Both machines are now logically protected against one machine outrunning the other in the same way that machine B waits in Figure 6.2.

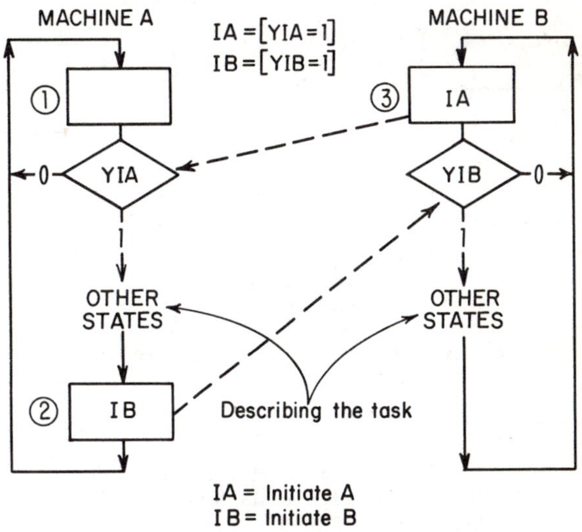

Figure 6.1 All ASM Chart Description of Two Serially Linked State Machines Using the Same Clock.

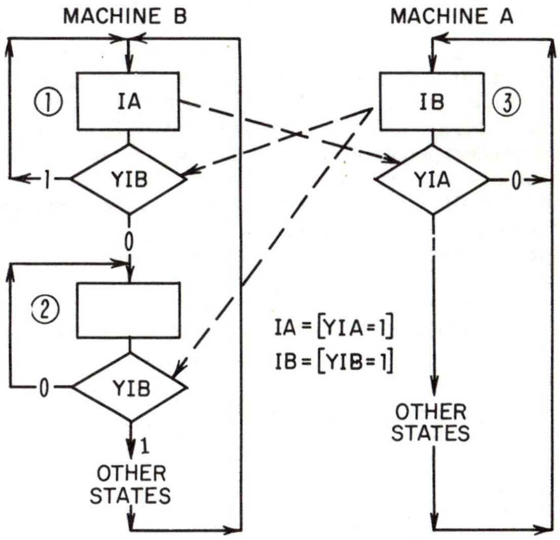

Figure 6.2 An ASM Chart Description of Two Serially Linked State Machines Where the B Clock Is Faster Than the A Clock.

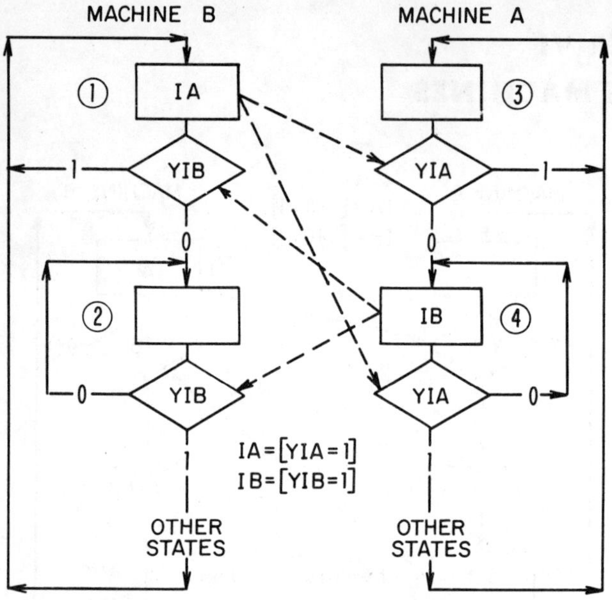

Figure 6.3 An ASM Chart Description of the Serial Linking of Two Machines of Unknown Relative Clock Times.

Parallel Linking

Two machines may also be linked parallel in time to provide parallel processing of information or decisions. Such machines are called PARALLEL LINKED. Figure 6.4 shows two machines, A and B, which work in parallel using the same clock. Both machines will start together because of the cross-linked qualifiers regardless of which machine finishes its cycle first. Again, when the B clock is faster than the A clock, the two machines must follow the description in Figure 6.5 to ensure correct operation. Figure 6.6 demonstrates how a third machine, C, must be added to act as the independent controlling element which enables two machines of unknown relative clock time to run in time parallel. This example completes the problems of interfacing two state machines with different clock rates in both parallel and serial linking. Each example demonstrates that a minimum number of states is required to accomplish certain types of interface. In each situation, these states must be designed into the instruments for the interface to work.

6.2 ITERATIVELY LINKED MACHINES

Machines that are linked but are in themselves a part of a chain of similar machines are called ITERATIVELY LINKED machines. Such configurations exist in a number of areas where the algorithm for the task is also iterative. Some of the common iteratively linked structures to be covered are adders, ripple counters and shift registers. Iterative linking is a special case of interface linking.

Adders

Many mathematic algorithms can be reduced to an iterative form which operates on each bit of some coded representation of a number in an identical manner. Binary addition is a typical example. A familiar representation of the sum of two numbers is the addition table. In Figure 6.7, two tables are shown, one for conventional decimal numbers and another for binary numbers. Below the diagonal stepped line in each table, the sum of the A and B digits is expressed with two digits. The additional 1 digit is called a CARRY because it represents a digit to be added to the next significant location. Obviously, binary addition is simple compared to decimal addition. An algorithm for adding two binary numbers involves the application of the addition table over and over again for each digit position. However, because the previous digit addition may have produced a carry, up to three digits need to be added in each position. Figure 6.8 gives a table for this addition and shows a circuit symbol to represent the combinatorial circuit performing this task. The circuit is called a FULL ADDER. The gate realization of a full adder can be found using Chapter 4.

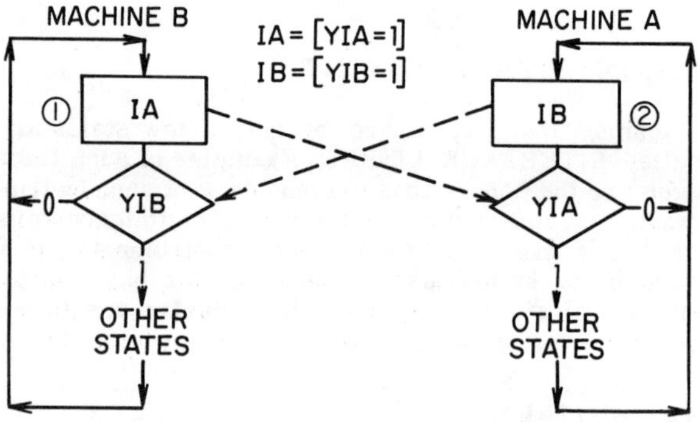

Figure 6.4 An ASM Chart Description of Two Parallel Linked State Machines Using the Same Clock.

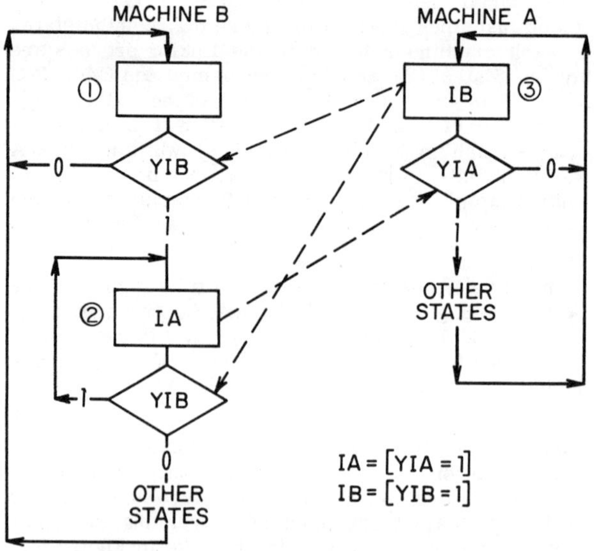

Figure 6.5 An ASM Chart Description of Two Parallel Linked State Machines Where the B Clock Is Faster Than the A Clock.

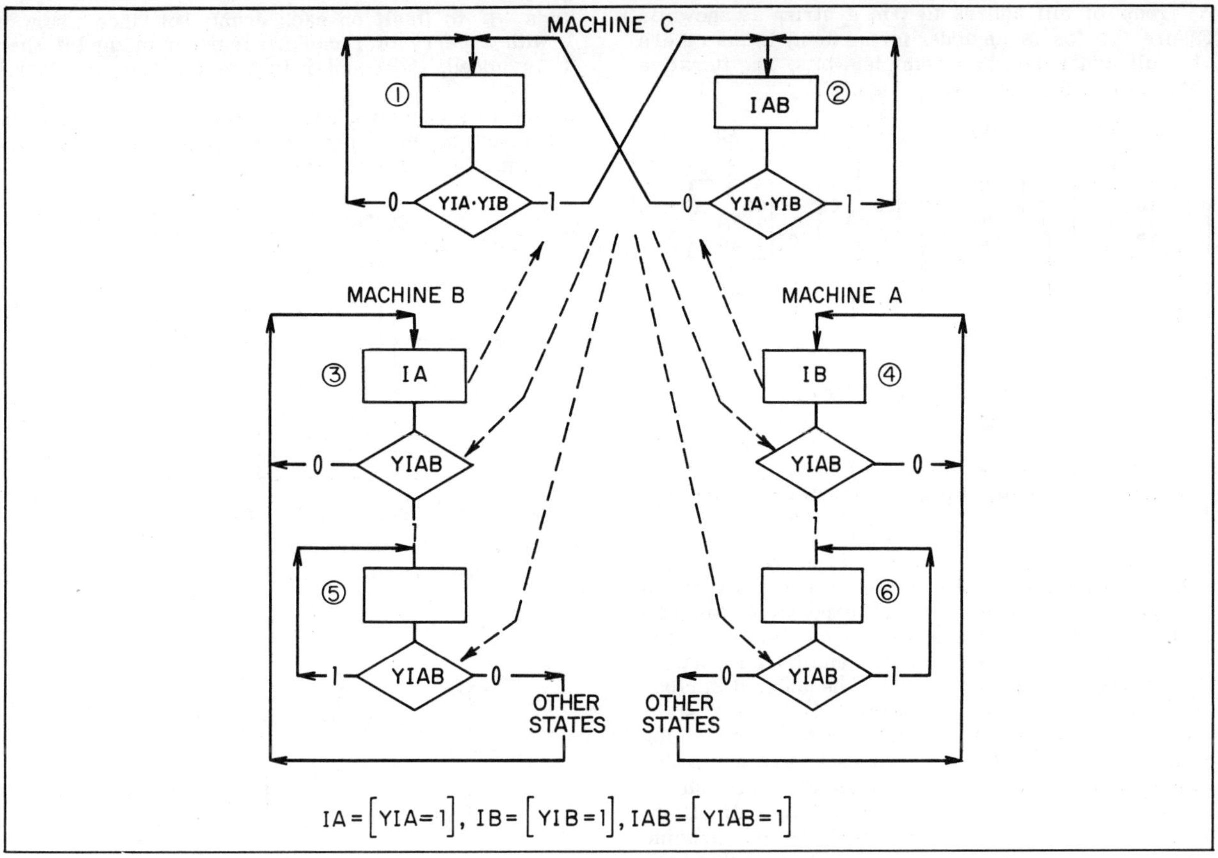

Figure 6.6 An ASM Chart Description of the Use of a Third Machine, C, to Define the Parallel Operation of Two Machines, A and B, of Unknown Relative Clock Times.

A \ B	0	1
0	0	1
1	1	10

BINARY ADDITION

A \ B	0	1	2	3	4	5	6	7	8	9
0	0	1	2	3	4	5	6	7	8	9
1	1	2	3	4	5	6	7	8	9	10
2	2	3	4	5	6	7	8	9	10	11
3	3	4	5	6	7	8	9	10	11	12
4	4	5	6	7	8	9	10	11	12	13
5	5	6	7	8	9	10	11	12	13	14
6	6	7	8	9	10	11	12	13	14	15
7	7	8	9	10	11	12	13	14	15	16
8	8	9	10	11	12	13	14	15	16	17
9	9	10	11	12	13	14	15	16	17	18

DECIMAL ADDITION

Figure 6.7 Decimal and Binary Addition Tables.

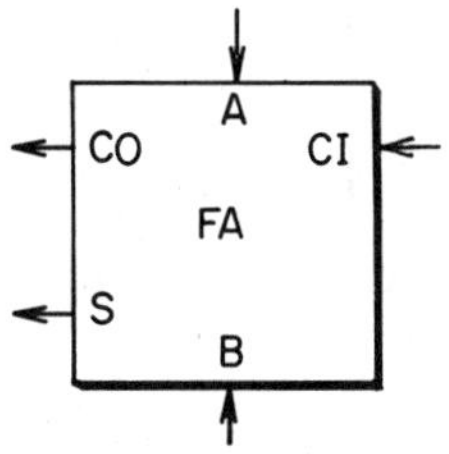

FULL ADDER

A	B	CI	CARRY CO	SUM S
0	0	0	0	0
0	0	1	0	1
0	1	0	0	1
0	1	1	1	0
1	0	0	0	1
1	0	1	1	0
1	1	0	1	0
1	1	1	1	1

Figure 6.8 Table Description of a Full Adder.

A group of full adders tied in a string as shown in Figure 6.9 forms an adder for as many bits as there are full adders, in this example 4 bits. The iterative character of the linking is evident.

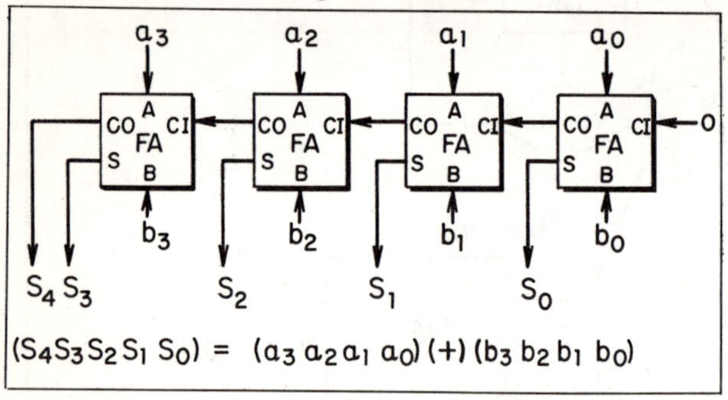

Figure 6.9 A 4-Bit Adder Made from Four Full Adders Showing the Iterative Structure of Linking.

Since the hardware for each bit of an iteratively linked structure is the same, a great hardware saving may be made at the expense of time by serially performing the operation on each bit using the gates for one bit and remembering the information for the next bit. The ASM chart for the adder described in Figure 6.8 could be represented by a 2-state machine which would remember the carry bit for each successive bit as shown in Figure 6.10. The gate implementation for this machine is also shown in Figure 6.11 to demonstrate the relative simplicity of the machine. Of course the equivalent machine could be made from one full adder and a delay flip-flop, as in Figure 6.12; but this solution is in reality more complicated in total gate count.

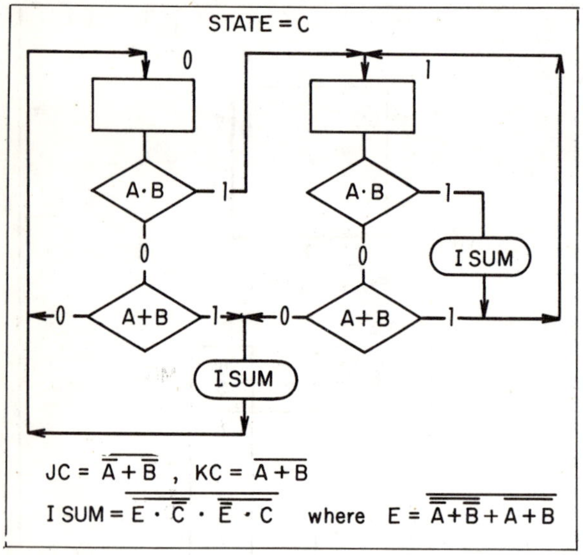

Figure 6.10 An ASM Chart of a Full Adder.

The other basic algorithms for subtract, multiply and divide also can be reduced to iterative linked structures. Some special adder functions can be implemented as an iteratively linked machine. For example, a counter is really an adder which always adds 1 to itself on each count; but since a sum of 1 with a carry of 1 can never occur in any bit when 1 is added, the counter is simpler than an adder.[1]

In previous chapters, counters were discussed as state machines, but now they will be considered as separate state machines linked together.

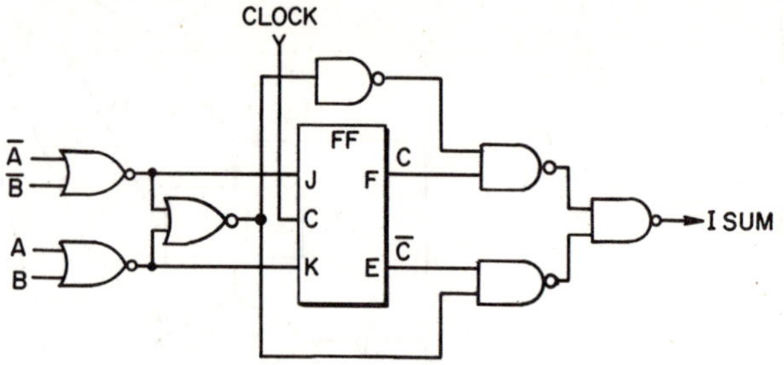

Figure 6.11 An Implementation of A Serial Adder and a D Flip-Flop.

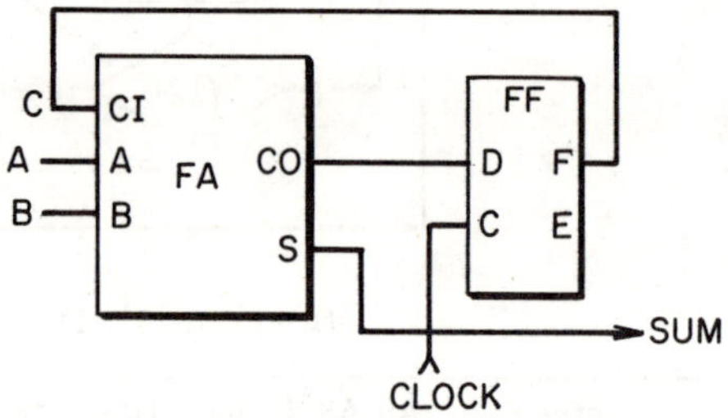

Figure 6.12 A Serial Adder Using a Full Adder and a D Flip-Flop.

Ripple Counters

A simple incrementing counter element will respond to a carry from a previous counter element. The least significant bit will always have a carry input to simulate the addition of a 1. Figure 6.13 describes an element of the counter under these simplifying conditions, assuming that the state time must be defined by the inputs alone. This ASM chart is similar to the description of a T flip-flop with T = 1 or a JK flip-flop with both J and K = 1. The carry input is analogous to the clock input, and a carry is indicated whenever the previous bit goes to a 0. Figure 6.14 then shows some typical elements of such a counter connected to count to 16. This type of counter is often called a RIPPLE COUNTER, which describes the progress of the carry propagation down the length of the counter. Since a carry may propagate down the entire length of the counter, going throug. the

---

[1] In binary addition, adding a 1 to a sum can produce a sum of 1 and no carry, or a sum of 0 and a carry but never both. In each bit position of the counter the situation is the same, caused either by the initial 1 to be added to the first bit or by the carry for the remaining bits.

delays in each stage, such a counter will not, in general, come to a stable count as fast as a counter designed as a single-state machine; but it has a simple circuit for binary counting.

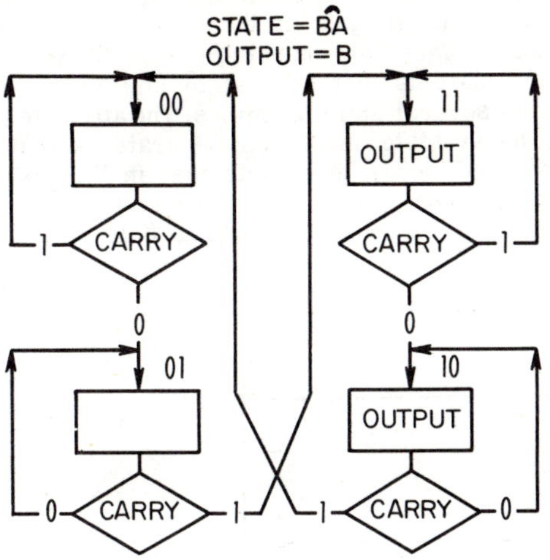

Figure 6.13 A Counter Element under the Assumption That the Input Controls the State Time.

Shift Registers

Another popular iteratively linked structure is the SHIFT REGISTER. The shift register is made from a string of delay elements using a common clock as described in Figure 6.15. The JK flip-flop could also be used to implement the shift register since a JK can be made to look like a D flip-flop with an additional inverted input from each previous stage.

Iteratively linked structures are generally slow in comparison with single-state machines. There are some modifications to such structures to speed them up that involve such features in an adder as LOOK-AHEAD CARRY to reduce carry propagation time. A look-ahead carry refers to additional combinatorial logic used to generate a carry more rapidly than it would be propagated through the adder.[2] However, for the most part, the significant characteristics of iteratively linked machines are their simple structure and their ability to be converted to a serial operator.

[2] For additional information see Herbert Hellerman, "Digital Computer System Principles," Section 7.9, McGraw-Hill Book Company, New York, 1967.

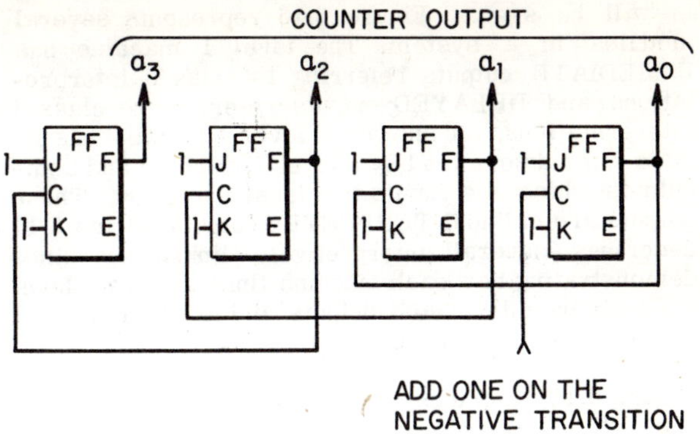

Figure 6.14 A 4-Bit Ripple Counter as an Example of a Special Add Circuit.

6.3 INTERPRETIVELY LINKED MACHINES

This section discusses a valuable concept useful in constructing machines to perform complex algorithms. This concept allows the design of machines that use other machines to become more powerful. The concept is called INTERPRETIVE LINKING and refers to the algorithmic relationship between the linked machines.

Two machines are called INTERPRETIVELY LINKED if each state of one machine can be described by the ASM chart of the other machine. In this way, one machine interprets the meaning of the other. The process of defining a particular array of machines was introduced in Chapter 2. This array or system is made from a number of functional modules linked together through a control module. The functional modules are described as immediate functions, class 0 machines, or as delayed functions, class 1 machines. The control module is usually a class 2 or higher machine. Each instruction of the control module, then, is interpreted by the other modules it controls. In a similar manner, the control just described may interpret another machine, and so on.

Levels and Total State

The relationship of one machine in a string of interpretive linkings is called the LEVEL of interpretation. A level 0 machine is either a class 0 or a class 1 state machine. Levels 1 and above are usually class 2, 3 or 4 state machines. The machines described in Chapter 4 and Chapter 5 were handled as level 0 and level 1 machines; however, the design techniques described in those chapters apply to all higher levels,

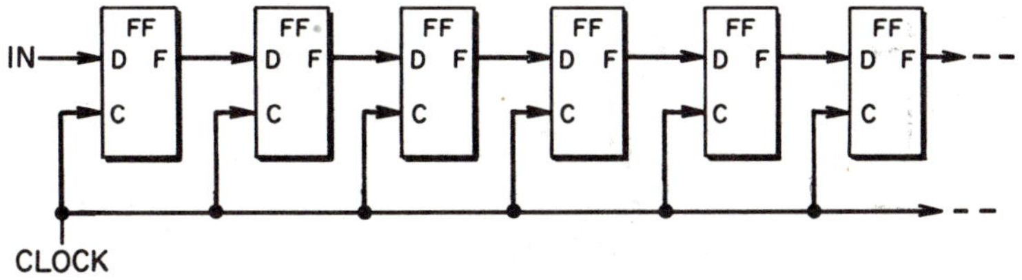

Figure 6.15 A Delay Flip-Flop Shift Register.

as will be seen. Figure 6.16 represents several modules in a system. The level 1 machine has IMMEDIATE outputs referring to class 0 interpretations and DELAYED outputs referring to class 1 interpretations, as shown. A level 2 machine has its state time determined by a level 1 machine and some outputs which are inputs to a level 1 machine. These outputs are called INTERPRETED outputs. Figure 6.17 describes several interpretively linked machines demonstrating how each machine links to a lower level machine by interpreted outputs and to a higher level machine by a state time output.

Figure 6.18 describes three interpretively linked machines to demonstrate the operation of levels and the concepts of total state. The TOTAL STATE is the concatenation of the states of all the levels. In an interpretively linked group of machines the total state establishes sufficient information to determine the next-state behavior just as the state does in a single machine. In fact, the only difference between total state and state is that total state implies a concept that spans several state machines. The attribute which makes these state machines separate is that they are designed separately. Note that in this example the output corresponds to the state so that the output behavior can be seen by looking only at the state behavior.

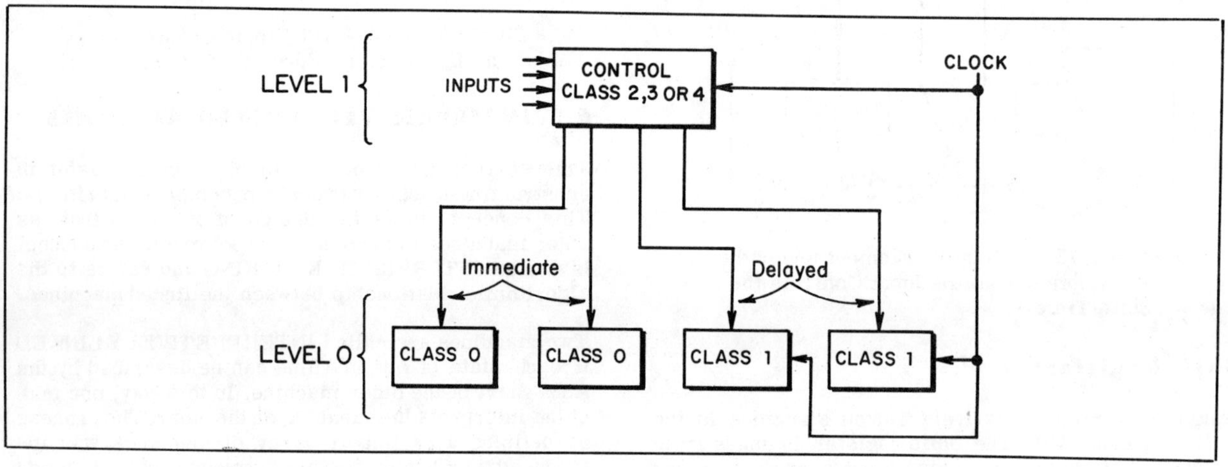

Figure 6.16 The Level Notation Applied to a Simple Two-Level System.

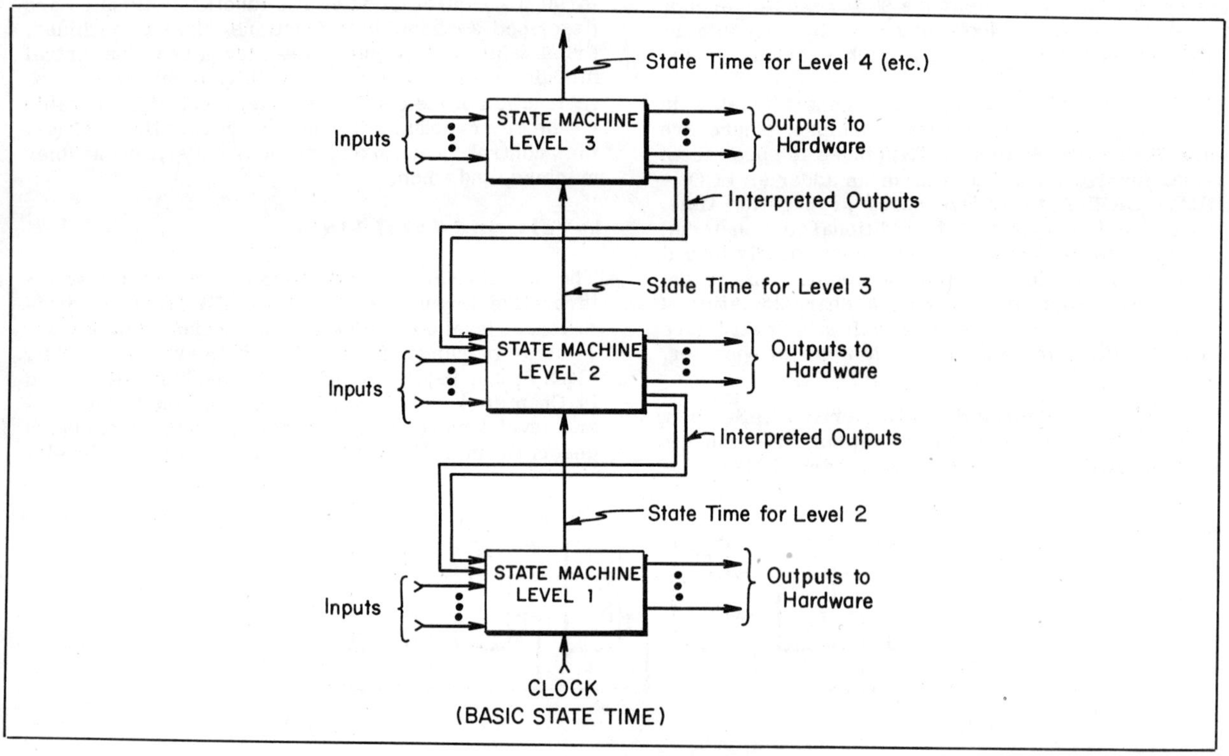

Figure 6.17 An Interpretively Linked Group of State Machines.

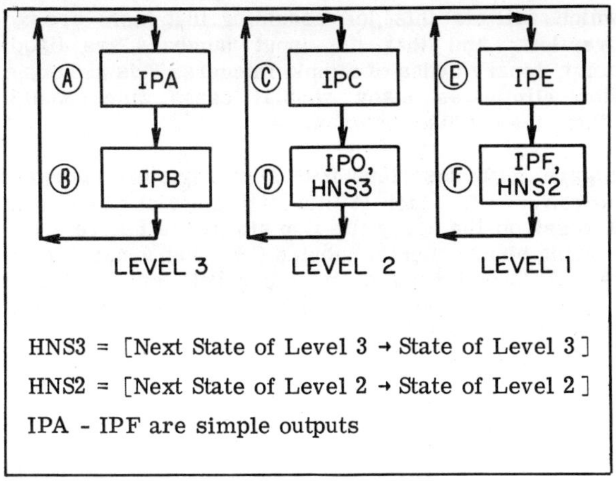

Figure 6.18 Three Interpretively Linked Machines.

Given that the initial total state of Figure 6.18 is given by (state of level 3)(state of level 2)(state of level 1) and is equal to ACE, the operation of the three machines proceeds as follows:

1. The level 1 machine, with a state time determined by a clock, steps to state Ⓕ. The new total state is ACF and the three outputs are IPA, IPC, and IPF.
2. At the end of state Ⓕ, HNS2 indexes level 2 to state Ⓓ and the new total state is ADE and the three outputs are IPA, IPD and IPE.
3. The clock advances level 1 to state Ⓕ making the new total state ADF.
4. At the end of state Ⓕ, HNS2 indexes level 2 back to state Ⓒ; however, HNS3 simultaneously indexes level 3 to state Ⓑ making the new total state equal to BCE.
5. In a similar manner the three machines continue through a sequence of total states which can be summarized from the initial state by the following list:

ACE
ACF
ADE
ADF
BCE
BCF
BDE
BDF
ACE
.
.
.

In terms of the state time of the level 1 machine, the outputs IPA, IPC and IPE can be traced in time as shown in Figure 6.19. The primary purpose of this example is to demonstrate the meaning of total state and the relationship of state outputs between levels. The highest level always has outputs of the longest duration as shown in Figure 6.19. The real power of interpretively linked machines comes in interpreted outputs to be discussed next.

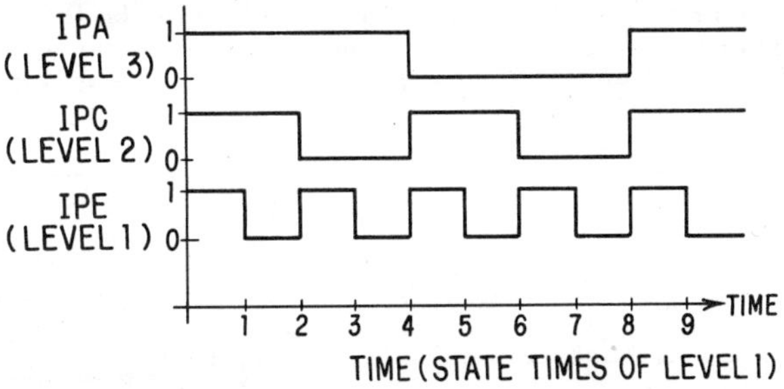

Figure 6.19 The Time Trace of Three Outputs from the Linked Machines Described in Figure 6.18.

6.3.2 Interpreted Outputs

Any instructions issued from a machine which are interpreted by a level 1 or greater machine are indicated by a star, *, preceding the mnemonic in the ASM chart. For purposes of simplicity, these instructions are called INTERPRETED INSTRUCTIONS. In the situation where all or many instructions are starred, an alternate, SPLIT, notation can be adopted, as shown in Figure 6.20.

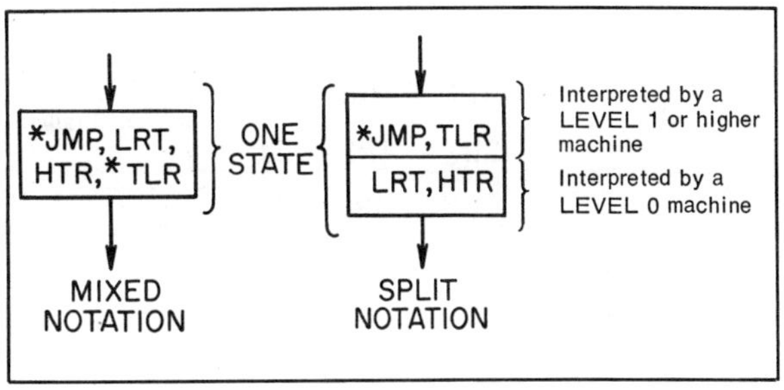

Figure 6.20 Notation for Interpreted Instructions in a State of an ASM Chart.

Instructions which are interpreted by a level 1 or higher machine may not be executed at the same time. This difference may cause some confusion in reading an ASM chart; therefore, where possible, the list of instructions is separated by slashes to indicate the order of execution.

For example, *JMP/*TDT means that JMP is executed, then TDT is executed. The prefix I is never used on an interpreted instruction, and the use of the prefix H or L is optional. The logic level is assumed to be H if H or L is absent.

The star, *, notation is used in Chapters 4 and 5 to denote an asynchronous qualifier. Actually, the star means that the qualifier is not defined in the time reference for the machine and that it is subject to further interpretation elsewhere. Thus, the star notation in general means that the definition of a signal is external to the machine being considered.

To illustrate the application of an interpreted linked pair of machines, a simple example of a serial machine will be used. This machine adds three numbers shifted in as an input and stores the sum in a register. Figure 6.21 describes a machine which will do this job assuming that there are no overflows and that the input numbers are fixed point, binary codes of simple integers. This assumption eliminates many special cases which would otherwise confuse the example.

Figure 6.21 describes four shift registers, labeled R0 through R3, each 16 bits long, a serial adder with a reset on the carry flip-flop similar to Figure 6.11, a 4-bit binary counter labeled CR, plus a number of terminals referred to by the labels on the block diagram. The two ASM charts describe two machines labeled M1 and M2.

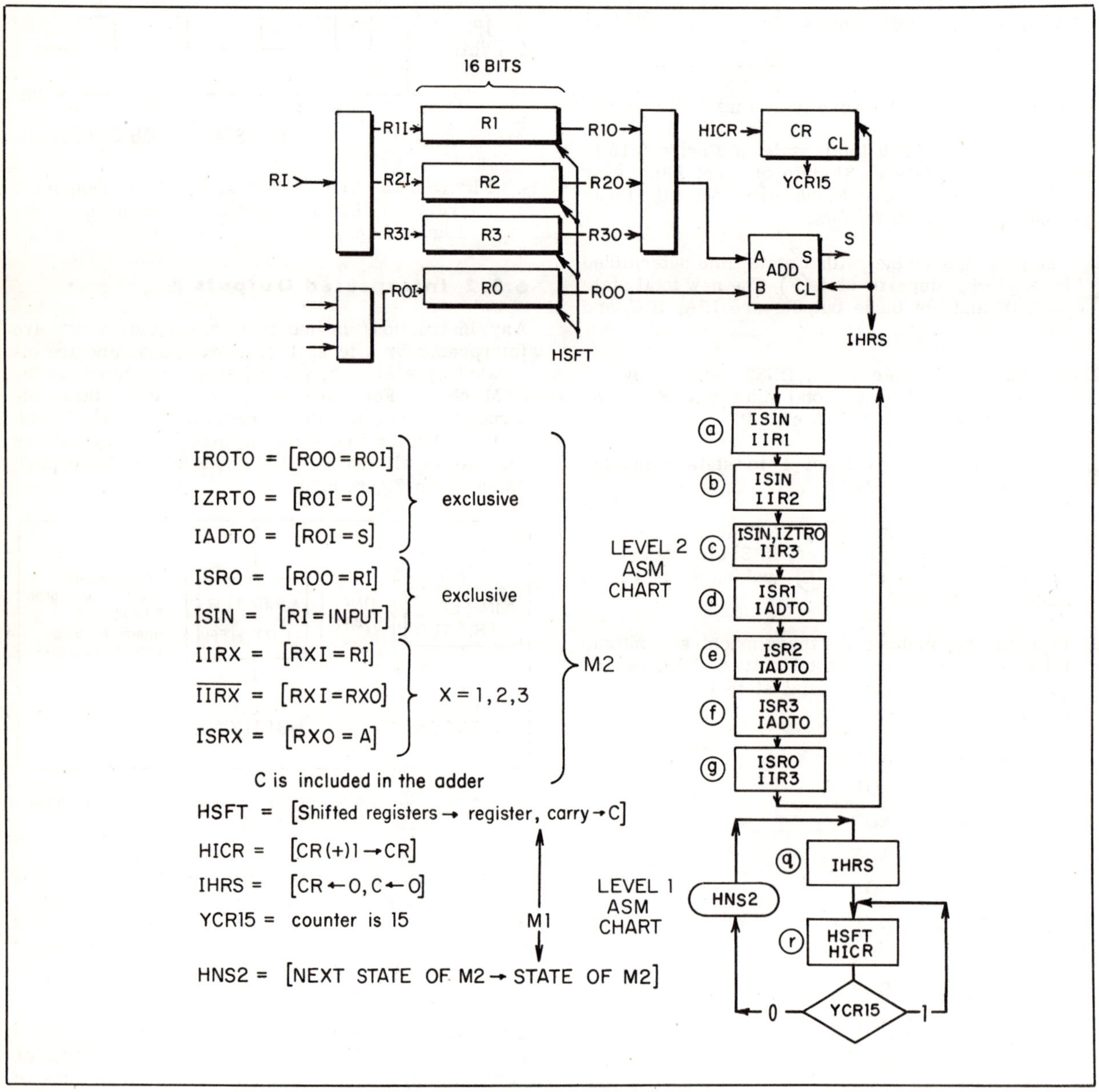

Figure 6.21 A Simple Machine to Demonstrate the Use of Interpretive Linking in Determining the State Time of a Level 2 Machine.

Machine M1 is level 1 and has three functions, which are: to shift the registers 16 positions to the right, to reset the counter and carry flip-flop, and to increment the state of the level 2 machine, M2. Machine M2 establishes the proper data paths in the machine in the proper sequence to perform the required job of adding three numbers from the input.

M1 starts in state (r) and M2 starts in state (a). M2 issues ISIN and IIR1 which connect the INPUT up to the input of RI. M1 shifts RI and increments the counter, CR, 16 times, ending up in state (q). The state of M2 is changed to (b) at the end of the 16th count by HNS2. State (q) resets CR and the flip-flop C. State (b) issues ISIN and IIR2 which connect the inputs to the input of R2. Meanwhile, M1 has gone to state (r) and starts shifting the next number into R2. Again, when 16 shifts have been made, M2 is set to state (c). It is easy to see that M1 makes every state of M2 perform an operation on one register length. Therefore, M2 can be considered separately as operating on registers. State (c) loads a number into R3 and loads 0 into R0. State (d) shifts R1 and R0 through the adder and puts the sum in R0. State (e) adds R2 and R0 and puts the sum in R0. State (f) adds R3 and R0 and puts the sum in R0. Finally, state (g) shifts R0 into R3. At the end of state (g), the sum of three successive numbers from the input is in register R3. Although this example has a lot of extra moves, it serves to demonstrate the interpretation of state time in machine M2 and the basic power of interpretive linking.

The previous example can be made more useful if the start of the acceptance of input data is controlled. An interpreted instruction, *ADD, can be used to start the process if a decision box and state are added after state (g) in M2, and an additional condition is added to M1 to interpret a wait instruction from M2.

Figure 6.22 shows the changes in the ASM charts to describe the interpretation of the instruction *ADD from a level 3 machine, M3, which will cause M2 to enter three numbers and add them as before and continue doing this sequence until ADD = 0. When *ADD = 0, M2 will wait in state (h) which issues *WAIT to M1 causing M1 to wait in state (q). Thus, with some simple ASM chart additions, a single logic signal, *ADD, has been defined as a series of shifts and adds that total 127 separate clock times of the M1 machine. The HADD in the BLACK JACK machine could be an interpreted instruction similar to *ADD.

The synthesis of interpretively linked machines is carried on in exactly the same manner as described in Chapters 4 and 5. Each level machine is a separate design. An interpreted instruction is another output that is the same as all other outputs of the state machine. Each level can be synthesized with gates or with ROMs just as before. The only difference is that the notation has been detailed to help explain the interpretive linking defining mnemonics which are outputs from one machine and inputs to another.

Computer Structures

It is instructive at this point to review briefly how the interpretive link structure pertains to three computer machine structures. It will become evident that the historical development of computers has merely followed a course of adding a new machine level for each new machine and, therefore, the structure of future machines may be extrapolated.

The most basic machine structure, the TURING MACHINE, is used extensively in automata theory to prove basic principles but is not really a practical computing machine. The Turing machine consists of an infinite tape, a head capable of marking and

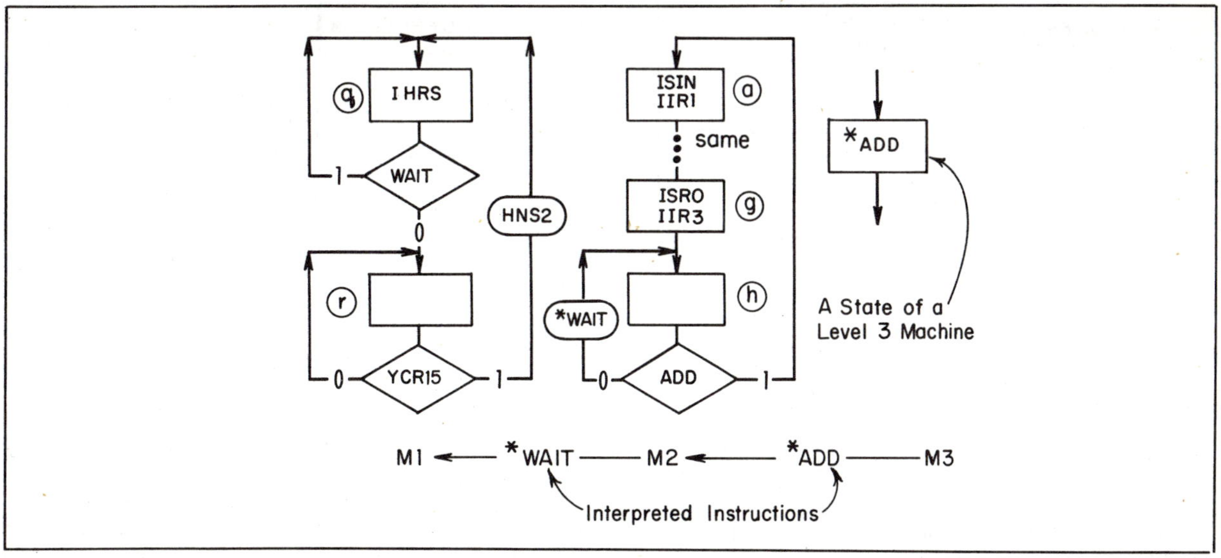

Figure 6.22 An ASM Chart Description of Two Interpreted Instructions *ADD and *WAIT.

reading the tape and a finite state control. The tape is divided into squares and the head can read and mark a finite number of symbols in this square.[3] The control can move the tape forward or backward one square at a time based on a decision made from the state of the control and the symbols read from the tape. The control may be represented by a level 1 ASM chart. The decision box structure has conditional immediate outputs to mark the tape and conditional delayed outputs to move the tape one way or the other. Figure 6.23 represents a possible ASM state from such a control and shows examples of mark outputs and move outputs.

Tape symbols S_i

IMS_i = mark S_i

IES_i = erase S_i

HLM = [Left square → square]

HRM = [Right square → square]

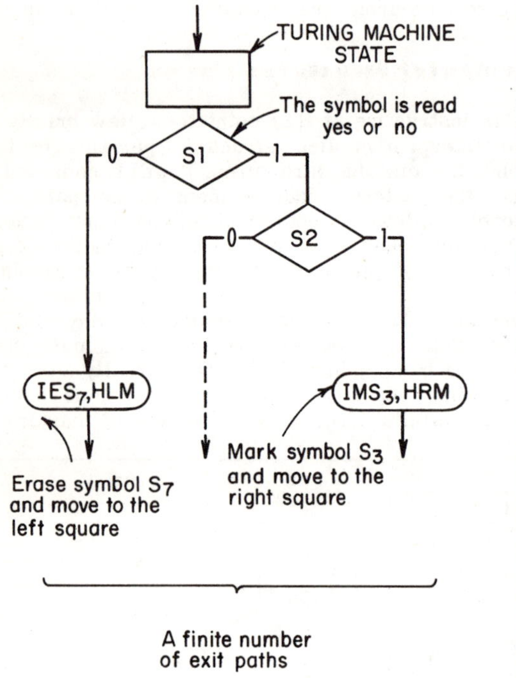

Figure 6.23 An Example ASM Block to Represent One Turing Machine Control State as a Level 1 Machine.

The second machine is the VON NEUMAN structure. This machine consists of a control, a memory and an arithmetic unit. The control is responsive to codes stored in the memory called the PROGRAM. Figure 6.24 describes a block diagram of these elements.

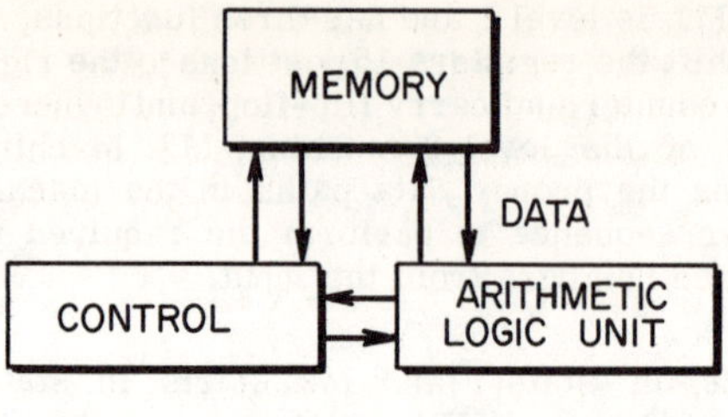

Figure 6.24 The Basic Von Neuman Machine.

The Von Neuman machine may be described as a level 2 machine. The arithmetic logic unit (ALU) and all the addressing of the memory are controlled by the control unit, which is a level 1 machine since the previous functions are immediate and delayed type. The control interprets portions of the code from the stored program, which therefore is a level 2 machine. As was seen in Chapter 5, a state machine can be implemented in a memory. Two ASM charts in principle, then, can describe any program stored in a Von Neuman machine.

The third type of machine is rather recent and is often called a MICROPROGRAMMED MACHINE. Figure 6.25 gives a block diagram impression of this machine showing that it is basically the same as a Von Neuman machine with one extra level of interpretation, called the MICROPROGRAM, between the stored program and the control. This machine is, therefore, a level 3 machine in the stored program, a level 2 machine in the microprogram, and a level 1 machine in the control. From these examples, it is easy to extrapolate that a machine could be built with any number of levels, each level being described by an ASM chart.

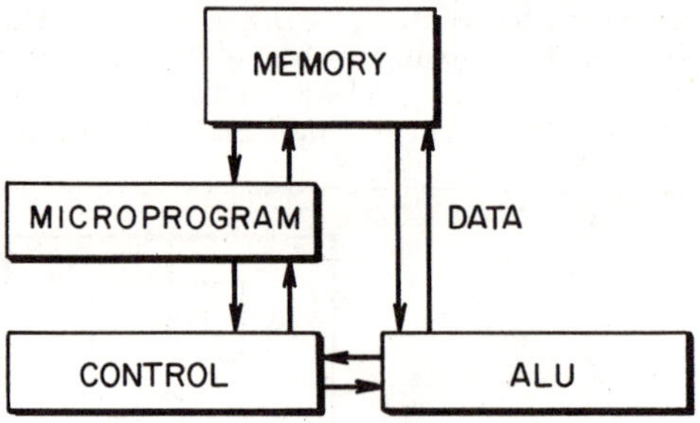

Figure 6.25 A Basic Microprogrammed Machine.

Interpretive linking has the potential for making machines perform far more complex operations than might be indicated by the sum of their individual states. The actual effectiveness is closer to the product of the state counts in the interpretive machines since each state of one level machine is defined by the states of the lower level machine. In reference to the discussion on memory addressing in Chapter 5, this statement indicates that all the levels should have approximately the same number of states for maximum utilization of a given number of states. Although this statement is only a guideline, it is interesting to look at the state count in the HP 9100A calculator: level 1 has 64 states, level 2 has 512 states, level 3 has 256 states and level 4 has 196 states. The three highest levels are very close in count and the first

[3] In its most primitive sense one head writes one symbol on one tape square, but there is no loss in generality by placing a finite number of tapes in parallel and thus allowing a finite number of symbols per square.

level is not really that far away. On the other hand consider a basic general purpose computer having two levels, such as the Von Neuman machine which may have seven states in the first level and 16,000 states in the second.

6.4 SOFTWARE LINKED MACHINES

The coding of a general purpose computer memory such that the computer performs a specific task is called PROGRAMMING the computer. The description of the coding is called SOFTWARE, primarily because it is on paper rather than in circuits. However, software and hardware are really implementations of the same type of logical task; both can be described by an ASM chart. The implication is that any software program can be implemented in hardware.

To illustrate the translation of some software statements into hardware, consider the ALGOL statement:

FOR I = 1 STEP 1 UNTIL 10 DO S_1 ;

The symbol S_1 stands for a statement to be interpreted. Figure 6.26 gives the ASM chart description of this same statement which may be converted to hardware by following Chapter 4 or 5. An interpretively linked machine can operate on S_1, telling the main program when to move to the next statement. The extension of this example to an entire software program could yield a complex machine. Still, the essential concept of ASM charts and algorithmic state machines is that once the task is adequately described, it can be implemented. A computer program is just another description of an algorithmic task.

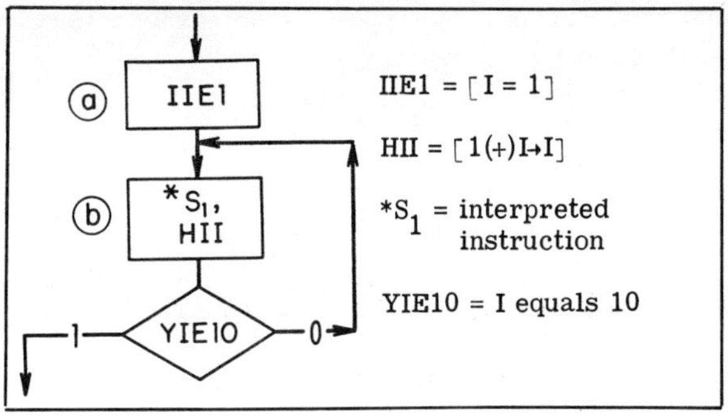

Figure 6.26 Implementation of the ALGOL Statement, FOR I = 1 STEP 1 UNTIL 10 DO S_1.

6.5 CONCLUSION

The linking of machines provides the power and flexibility to perform complex algorithmic tasks. From the early discussions of functional partioning in Chapter 1, the discussion of algorithmic state machines has been carried through description and design and finally back to the problems of partitioning a structure of linked machines. Interface linking defines the operation of two simply linked machines. Iterative linking introduces some basic types of iterative algorithm implementation. Interpretive linking introduces the concepts of interpretively defined outputs. Interpretively linked ASM charts can be used to describe complex machine structures, such as a microprogrammed computer, almost as easily as an ASM chart can describe a JK flip-flop. Finally, the computer programming languages may be represented as an interpretively linked group of algorithmic state machines.

REFERENCES

CHAPTER VI

Bell, G. et. al.: "A New Architecture for Mini-Computers -- THE DEC PDP-11," AFIPS Conf. Proc., Vol. 36 (Montvale, New Jersey: AFIPS Press, 1970), pp. 657-675.

Hellerman, Herbert: Digital Computer Systems Principles (New York: McGraw-Hill Book Company, 1967).

Husson, Samir S.: Microprogramming Principles and Practice (Englewood Cliffs, New Jersey: Prentice-Hall, Inc., 1970).

Ramamoorthy, C. V.: A Study of User-Microprogrammable Computers," AFIPS Conf. Proc., Vol. 36 (Montvale, New Jersey: AFIPS Press, 1970), pp. 165-182.

CHAPTER VII
INTRODUCTION TO PERFORMANCE EVALUATION

7.0 PERFORMANCE EVALUATION

Simulation and test are introduced in this chapter as two means for evaluation of a logic design before and after hardware is constructed. Specific discussion covers some design considerations to make a state machine testable.

7.1 SIMULATION

The simulation process is described in this section. The use of a SIMULATION LANGUAGE is introduced as part of a system called the SIMULATOR. The details of such a language are omitted. A parallel is drawn between the INTERPRETER and hardware test.

The process of SIMULATION involves creating a make-believe machine which operates according to the rules established by the description, such as an ASM chart, and by the definitions, such as terminal definitions. The most common simulation is done manually, on paper or in the mind. This process necessarily accompanies the development of an algorithm from the kernel of an idea. Manual simulation suffers from slowness and inaccuracy. Since the designer knows what he wants to do, he often smooths over little errors while simulating a machine. Later, these errors cause problems in the hardware. The less common but more effective simulation is done with the aid of a computer. The computer does exactly what it is told and therefore checks the complete accuracy of the description and the definitions. While a successful simulation guarantees that the algorithmic process is correct, it does not check the electrical circuits which form the machine and implement the task. It only checks their logic definitions. The study of electrical circuits is a separate design problem.

The computer simulates an algorithmic state machine by copying the performance of each state in the description. As in the model, the state time is divided into a stable and a transition period, and the state is defined during the stable part. In this way, the computer processes the state description according to the appropriate definitions in any manner that is convenient to produce the correct result in the next stable period. The actual time of the simulation is unimportant for a logical check, although a time reference tally can be maintained to accumulate the time used by the make-believe machine. Just as in the real machine, the contents of the memory elements during the stable portion of the state time determine the operation in the next state.

The simulation on the computer requires changing the machine description and definitions into a form acceptable to the computer. The computer accepts character strings. This change is achieved by translating the definition statements, the block diagram connections, the equations, the tables, the maps and the ASM charts into one language called the SIMULATOR LANGUAGE. The simulator language is constructed to make this transition easy and represents all descriptions as character strings.

The simulator language is fed into a computer which is programmed to read and organize the information into a listing of codes which represent interpreting instructions and symbol definitions. This listing is then interpreted once again by another program, called the INTERPRETER, which provides the actual means to execute the information in the listing and update the memory elements for each new state of the make-believe machine. The interpreter has features which provide the means for evaluation of the performance of the make-believe machine through outputs to a

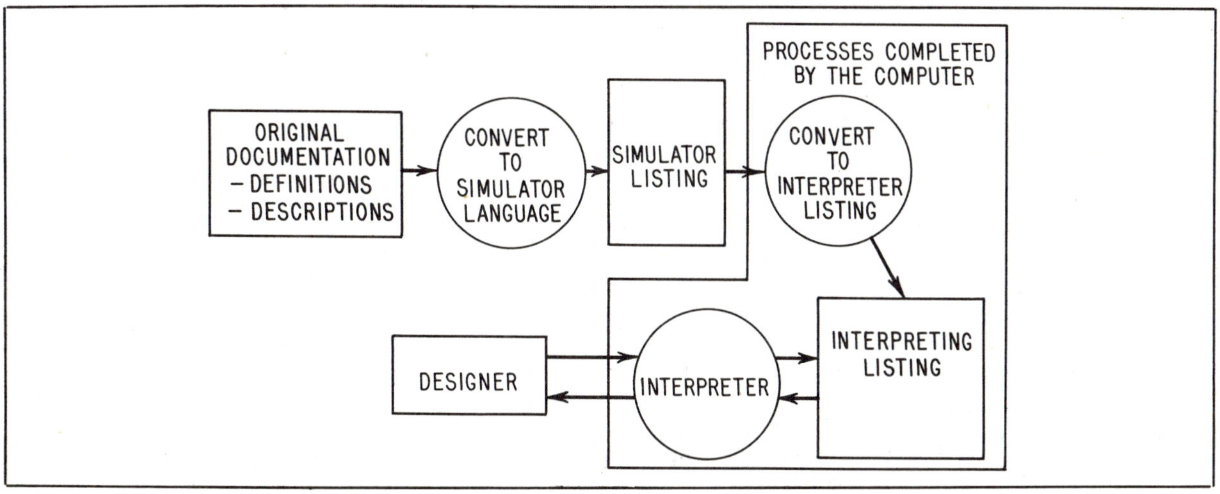

Figure 7.1 The Computer Simulation Process.

printer or visual display. The designer can alter, at will, the type of information he wants to see during the course of the simulation. Figure 7.1 describes the processes involved in using a computer to simulate a state machine. In this diagram, the circles represent a process and the rectangles represent a result.

The interpreter provides access to the make-believe machine being simulated in a number of ways.

1. Selected memory elements may be viewed at each state time. These elements are selected by PRINT commands.

2. The make-believe machine can be made to execute one state at a time or run until some preset condition is met. Stopping at each state is called ABSOLUTE HALT. Stopping only on a condition is called CONDITIONAL HALT. Absolute and conditional halts can be set up for each level of an interpreted machine. This feature allows checking the output against the desired output for each state time.

3. Inputs can be selected by the designer to test alternate paths based on the inputs. This feature allows checking interface responses.

By using these three features, the make-believe machine can be run and logically checked against the desired responses.

Figure 7.2 shows a possible computer output for a portion of a machine, showing the results of several machine states by the printouts of the state code and the registers R, S, T and B. At each state, the contents of the registers are printed as they would appear during the stable portion of the state time. Comparing the printout with the ASM chart demonstrates the usefulness of the checkout simulation.

The simulation process provides a simplification when modules of the logic machine have been designed or are only defined. For example, a serial processing module may produce an arithmetic result upon two parallel inputs. To check out a system which uses this module, the transform equivalent to the module can be used so that the details of the serial arithmetic process need not be simulated or described. This replacement can be made for any module which can be defined over some time interval as a class 0 or a class 1 machine. The simulator can also provide simulation with states identified by symbolic notation rather than codes. The states can be simulated under these conditions because the state sequence is still well described. State assignments can be made and the equations for the next-state function can be checked after the algorithm has been verified using symbolic states. In this way, a partial separation of the state sequence simulation and the output function simulation is achieved. Thus, simulation is possible at several levels involving block operations and state assignments.

Simulation makes changes easy to check. The change is made in the simulator listing, and the computer makes the appropriate changes in the interpreting listing. The interpreter then allows convenient access to check the validity of the change.

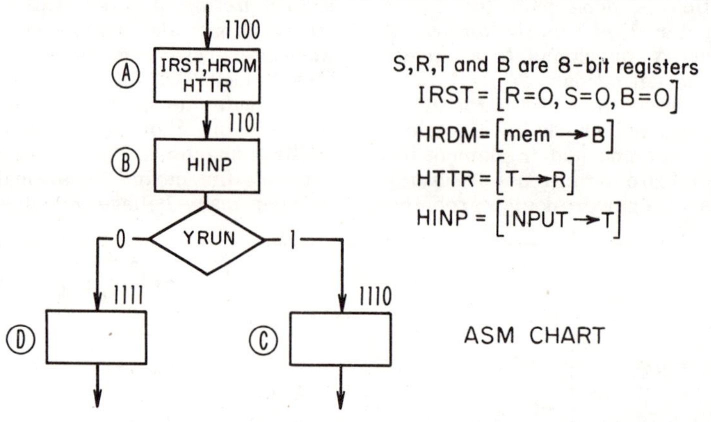

SIMULATOR OUTPUT

```
STATE=A, CODE=1100, R=00000000, S=00000000, T=00001111,
         B=00000000

STATE=B, CODE=1101, R=00001111, S=00000000, T=00001111,
         B=00101011
YRUN= 1
INPUT= 01010011

STATE=C, CODE=1110, R=00001111, S=00000000, T=01010011,
         B=00101011
```

Figure 7.2 An Example Simulator Output

Simulation can provide other outputs for testing the hardware. Since the simulator creates a make-believe machine which behaves like the real one as far as the logic is concerned, the interpreter could generate some test routines for the real machine. Now, the test phase is considered.

7.2 TEST

The process of test involves the logical and electrical check of a real machine. Much in the same way as in simulation, the test process uses a TESTER which provides the same access to the real machine that the interpreter provided to the make-believe machine. This section describes such a test system.

The TESTER is a device consisting of an octopus-like set of input-output lines, a control interpreter and an operator display console. Figure 7.3 describes such a general logic tester. The display provides meaningful data to the operator. The data is in the form of logic definitions. The analog information has to be obtained by direct connection to the devices or system under test. The control-interpreter takes the logical inputs from the operator console and makes the correct connections to the logic under test. A control interpreter may be very simple when inputs have to be provided. For example, one switch can be used for each input line and one light for each output line. The interpretation takes place in the labels on the switches and lights. These labels tell the operator the meaning of the controls. The control interpreter must also provide a means for single stepping each level of the machine. Figure 7.4 shows what a tester panel might look like when made specifically for the Black Jack machine. There are three major areas on a tester panel display: the input-output controls and display, the internal states display and the state time controls. These three areas are outlined in Figure 7.4.

The tester can be custom-built for each device tested, or a general tester can be programmed to accept the device and interpret the controls correctly. This general tester may be programmed by the simulator when it simulates the machine. The program would include a listing of which input-output wires would be tied to which logical terminals.

Extending the testing process a little further, the tester and the simulator may be combined. The simulation process can be compared to the response of the real machine and any differences noted on an output. Eventually, it might be possible to let the computer do some initial troubleshooting to isolate the problem of making correction easier. This is the real power of computer testing of real machines.

Some real machine problems may be caused by linear circuit problems. In these situations, the tester could be used to repeat the troublesome period over and over so that an oscilloscope could be used to study the linear signals.

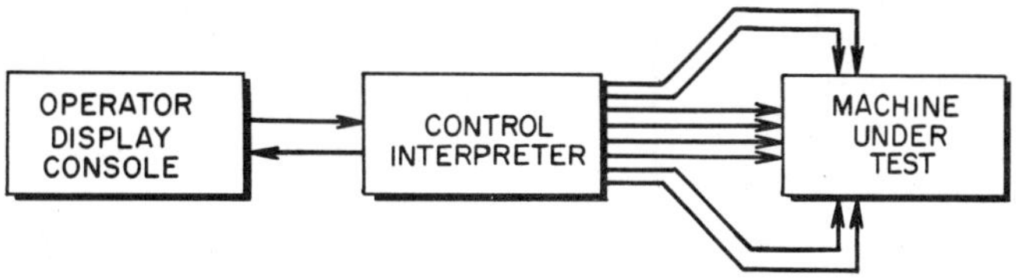

Figure 7.3 Basic Logic Tester.

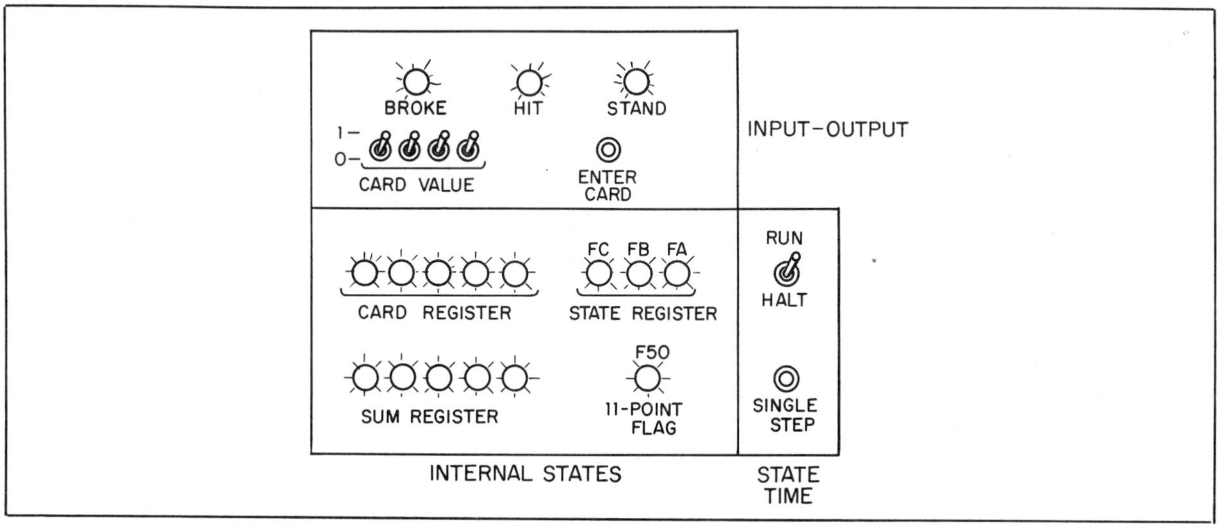

Figure 7.4 A Possible Tester Control Display for the Black Jack Machine.

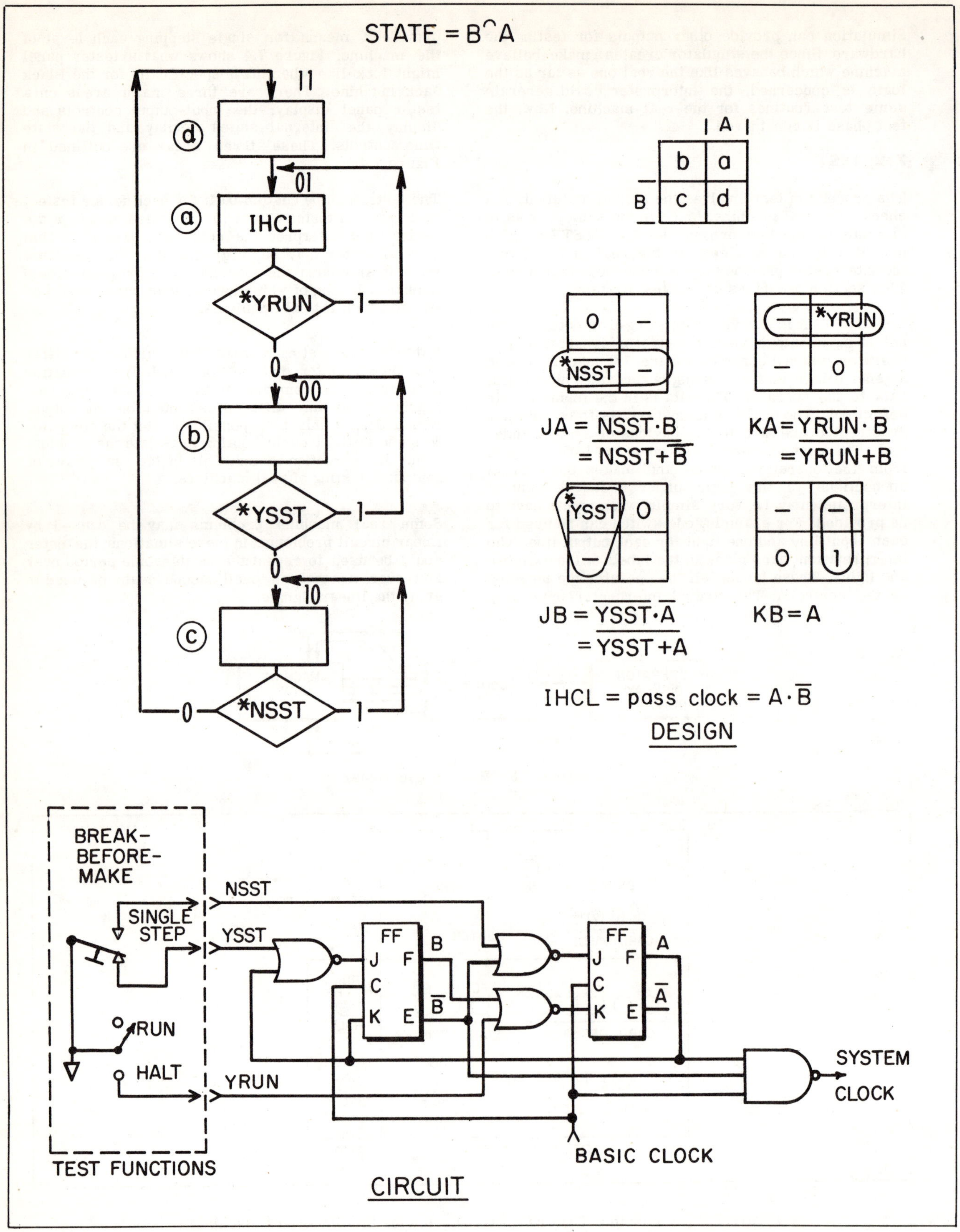

Figure 7.5 A Circuit Demonstrating the Use of Run, Halt, and Single Step in the Design of a Testable Clock Generator.

Thus, analogies can be drawn between simulation and test in that they are both performance evaluations of a real algorithm. The test is made on a real machine rather than a make-believe one. The tester provides the means to interpret the logical significance of the signals in the machine. Accordingly, a tester may be simple or complex, depending on the amount of interpretation required by the operator.

7.3 DESIGN FOR TEST

In order to provide a meaningful test of a system, certain design features must be included into the design of a system. Other features are also useful for self-test but are not required for the initial test. Both of these areas are touched upon in this section.

Each level of a logic system should be designed to be single stepped. This feature requires that states and qualifiers be added to the state machine specifically for this purpose. For example, Figure 7.5 describes a portion of a state machine interpreting a higher level machine. This machine, in fact, is the clock generator for a level 1 machine which can be logically controlled for RUN and SINGLE STEP by simple switch connections to the circuit. No alterations to the circuit or special pulse generators are required. Figure 7.6 shows a very similar ASM chart description of the single step and run feature designed into a more complicated machine. This feature should be included in each level of an interpretively linked structure of modules. Both of these examples use a single line, called YRUN, to determine if the state machine is to run normally or if it is to be single stepped. The single step button generates YSST and NSST. Two signals are used to prevent races caused by contact bounce on the switch. The switch is a break-before-make, which means that during the process of pushing the single step button, NSST first goes to a 1, then YSST goes to a 0. This progression prevents the machine from passing states ⓓ and ⓐ more than once for each push of the button.

Self-test features can also be designed into the logic. For example, the ASM chart in Figure 7.7 describes a possible test sequence for checking a read-write memory. The test sequence provides a repeated read-write on a single memory location with a test for comparison of read and write and optional incrementing of an address counter. If the comparison fails, the test halts with the address of the failing location. Such a test is applicable to core memories, which require a separate read and write signal. Self-test routines should be included in interpretively linked machines to provide a minimum cycle for the next level. In this way, a portion of the system can be checked relatively independently from the rest of the system.

The philosophy behind machines with self-test is to provide convenient logical isolation between major modules of a system and to support the isolation with rudimentary test sequences which exercise the isolated module. This philosophy of self-test depends upon a design founded on a sound logical organization of the machine. Such organizations are studied in the field of computer architecture. The study of machine organization is supported by the study of algorithmic state machines, which this book has attempted to describe in detail. Algorithmic state machines, therefore, aid in the broad concepts of organization, implementation and evaluation.

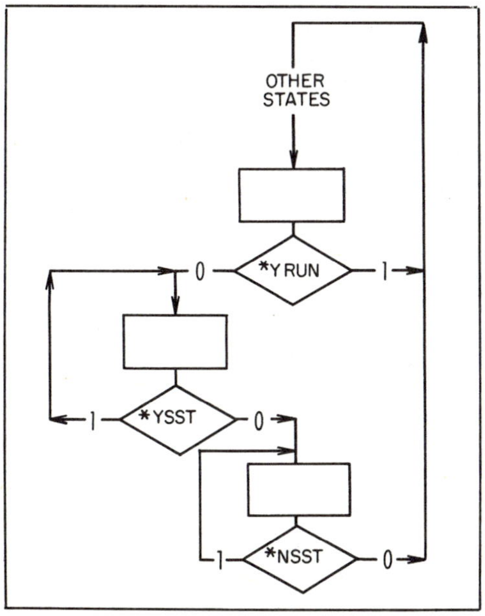

Figure 7.6 Run and Single Step Appended to a State Machine Description.

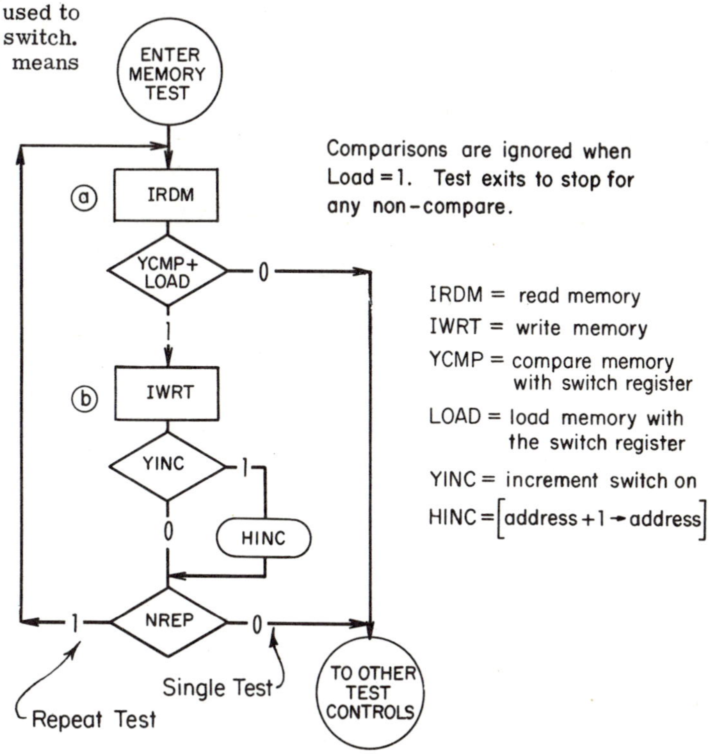

Figure 7.7 A Memory Test Sequence.

REFERENCES

CHAPTER VII

Bartow, Neil, and McGuire, Robert: "System/360 Model 85 Microdiagnostics," AFIPS Conf. Proc., Vol. 36 (Montvale, New Jersey: AFIPS Press, 1970) pp. 191-196.

INDEX

INDEX (Cont'd)